Explore Evolution

The Arguments For and Against Neo-Darwinism

Stephen C. Meyer, Scott Minnich,
Jonathan Moneymaker, Paul A. Nelson,
and Ralph Seelke

Hill House Publishers™
Melbourne & London

www.hillhouse–publishers.com
First published in 2007
by Hill House Publishers TM (Melbourne & London)
c/o O'Brien & Partners
Suite 4, 91 Station Street
Malvern, Victoria 3144, Australia

www.exploreevolution.com

Copyright © Stephen C. Meyer
Cover photo composite © Brian Gage Design

978-0-947352-47-9
ISBN: 0-947352-47-3 (Hard bound)
978-0-947352-48-6
ISBN: 0-947352-48-1 (Limp bound)
The Cataloging-in-Publication Data is on file at the Library of Congress.

Editorial: Lucilla Wyborn d'Abrera, London, UK
Art Direction/Design: Brian Gage Design, Vancouver, USA
Design: Calvin Carl, Elena Wilken, and Marc Niedlinger, Vancouver, USA
Design consultant: Derrick I Stone Design, Melbourne, Australia
Production by A-R Bookbuilders, Singapore
Printed and bound in Singapore
All rights reserved. No part of this publication may be reproduced, stored in a retrieval system, or transmitted in any form or by any means, electronic, mechanical, photocopying, recording or otherwise, without the written permission of the publisher.

Table of Contents

Preface

It's been nearly 150 years since the publication of Charles Darwin's influential book, *On the Origin of Species*, and the theory of evolution remains the focus of intense public controversy. So, what's all the controversy about?

It's hard to tell, sometimes. Public debate over this issue usually generates more heat than light. There is no shortage of headline grabbers willing to "go public" with confrontational statements on one side or the other. The media sometimes makes the situation worse. Television news in particular tends to rely on "sound bite" coverage of sharply divided opinions that are easy to describe. This makes for exciting viewing, but is not always helpful in finding answers to the real questions in the origins debate.

This focus on media-ready "good guys" and "bad guys" completely misses some of the real (and more interesting) scientific controversies about evolution. We hope this book will help you understand what contemporary Darwinian theory is, why many scientists find it persuasive, and why other scientists question key aspects of it.

Actually, we shouldn't refer to the theory as "it." As you will discover in the pages that follow, Darwin's theory is not a *single* idea. Instead, it is made up of several related ideas, each supported by specific arguments. We hope to help you understand what these ideas and arguments are and what different scientists have to say about them in light of current scientific evidence.

The approach we are using in this book is called "inquiry-based" education. This approach allows you, the student, to follow the process of discovery, deliberation, and argument that scientists use to form their theories. It allows you to evaluate answers to scientific questions on your own and form your own conclusions. Our goal in using this approach is to expose you to the discoveries, evidence, and arguments that are shaping the current debates over the modern version of Darwin's theory, and to encourage you to think deeply and critically about them.

Why use the inquiry-based approach?

The inquiry-based approach to education has a number of advantages. First, by enabling you to think critically about scientific theories and ideas, the inquiry-based approach will prepare you to be a better, more informed citizen. You will soon be asked to decide on many political and personal issues that involve science—debates about stem-cell research, decisions about personal medical care, and issues of environmental policy. Teaching scientific ideas openly and critically not only helps prepare you for possible careers in science, but it helps you learn to make informed decisions about such issues.

The second advantage to inquiry-based education is that students typically enjoy science more when it's taught this way. Scientific conclusions don't just pop up fully-formed from a lifeless collection of facts, so why would we teach science that way? Instead, the inquiry-based approach teaches about the arguments scientists *have* had, and *are* having, about current theories in light of the evidence. This allows you to do what scientists do—think and argue about how best to interpret evidence.

Third, many science educators are convinced that students gain a better understanding of a subject if they are taught about the arguments that scientists have in the process of formulating their theories. For this reason, the educational standards of several countries now encourage this approach.

United States federal education policy, for example, calls for teaching students about competing views of controversial scientific issues. As the U.S. Congress has stated, "[W]here topics are taught that may generate controversy (such as biological evolution), the curriculum should help students to understand the full range of views that exist."* In the United Kingdom, the National Curriculum for Key Stage 4 Science now recommends that, "Pupils should be taught how scientific controversies can arise from different ways of interpreting empirical evidence (for example, Darwin's theory of evolution)."

Controversies in science are nothing new. As recently as the early 1960s, for example, most geologists accepted the "geosynclinal theory" as the explanation of how mountain ranges form. After a significant period of controversy, most scientists came to accept the theory of plate tectonics because it provided a better explanation for a larger number of scientific observations. Yet without understanding the arguments that led to the acceptance of plate tectonics, it is very difficult to understand the theory itself or its current standing in the scientific community.

Today we continue to have important unresolved scientific controversies in many branches of science. In climatology, for example, scientists disagree over what global warming is, whether it is a natural phenomenon or a man-made problem, how big a problem it presents, and what (if anything) should be done about it. In theoretical physics, scientists disagree over the meaning and importance of string theory.

This book is one of the first textbooks ever to use the inquiry-based approach to teach modern evolutionary theory. It does so by examining the current evidence and arguments for and against the key ideas of modern Darwinian theory. We hope examining the evidence and arguments in this book will give you a deeper understanding of the theory and help you to evaluate its current status.

But are there really any *scientific* controversies about the modern theory of evolution?

Throughout this book, you will discover that there are, indeed, important scientific controversies about the key claims of evolutionary theory and about the arguments that are used to support them. We have written this book, in part, so that you could learn about the controversial aspects of evolutionary theory that are discussed openly in scientific books and journals but which are not widely reported in textbooks.

As we said before, Darwin's theory is made up of several ideas, each with supporting arguments. For each argument in Darwin's case, we will begin by explaining the argument, and examining the evidence in support of it. (We call this the Case For.) Then, we will spend some time examining the claims and evidence that lead some scientists to question the argument. (We call this the Reply.) We then look at the current state of the discussion in a section called "Further Debate."

Throughout the book, you may notice that the Reply section is often longer than the Case For section. There is an important reason for this. The Case For is the version taught in most school textbooks, and you should, therefore, already be familiar with it to some extent. The Reply section has not yet been presented in most school textbooks. The Reply is sometimes longer simply because it often takes more time to explain an unfamiliar concept or idea.

---

* This statement occurs in the authoritative conference report language of the No Child Left Behind federal education act.

Whenever there is disagreement over a particular point, we have tried to give arguments from the "best" people we could find on both sides of the question, rounding up the most qualified proponents and critics that we could.

Ah, yes: the critics. Who are these people, anyway? The main thing you need to know is that "the critics" are not necessarily the same from chapter to chapter. A scientist quoted in the Case For section of one chapter may very well be quoted as a critic two chapters later. As you will find out throughout this book, there are qualified, respected scientists on both sides of each argument.

Finally, you should know something about us, the authors. Two of us are biology professors doing research on evolution-related topics. Two of us are philosophers of science who have specialized in studying the logic of evolutionary arguments. One of us is a science curriculum writer. All of us happen to have reservations about various aspects of contemporary evolutionary theory, but we all think that students should learn *more*—not less—about this theory than they presently do. So, while we present criticisms of the theory that many biology books don't present, we also explain and develop the arguments *for* contemporary Darwinian theory in more detail than other standard textbooks.

One final word. We don't want you to simply accept this book as the last word on this subject any more than we'd want you to uncritically accept the word of other textbooks that present only the case for Darwinian evolution. That's the beauty of open inquiry—and of science, itself. That's also an example of the kind of critical thinking that we hope this book will encourage. Look at the evidence, listen to the arguments, and think for yourself.

So, what's the story on evolution? That's what we want to explore. What do you say we get started?

We also invite you to check out **www.exploreevolution.com** periodically for more information on these topics.

INTRODUCTION

Introduction

The modern theory of evolution has its roots in Charles Darwin's 1859 book, ***On the Origin of Species by Means of Natural Selection.**** *But nearly 150 years after the publication of the* ***Origin****, the theory of evolution remains the focus of intense public controversy. There are many reasons for this, of course: some philosophical, others political, or even theological.*

The purpose of this book, however, is to examine the *scientific* controversy about Darwin's theory, and in particular, the contemporary version of the theory known as neo-Darwinism. We hope to help you understand what this theory is, why many scientists find it persuasive, and why other scientists question the theory or some key aspects of it.

Historical Science: How We Explain the Past

All theories of origins confront us with the challenge of explaining the unobservable past. These theories try to explain unseen events, such as the origin of plants and animals—or the origin of our own species, *Homo sapiens*. This task can be difficult because, for nearly all of the history of life on Earth, there was no one to observe these events. Fortunately, this lack of eyewitnesses doesn't mean that we can't study life's history. It just means we have to use a different strategy.

Scientists who try to reconstruct the past—even the relatively recent past—must use different methods from those used by scientists working in a laboratory. *Experimental* scientists can observe phenomena under controlled conditions. However, *historical* scientists, like archaeologists and paleontologists, must try to figure out what happened in the past without the benefit of observing the past directly. They operate more like detectives, observing and studying the clues left behind by past events. Then, they try to reconstruct nature's history using what they know about the evidence, and what they know about cause-and-effect relationships. As the late Harvard paleontologist Stephen Jay Gould once explained, historical scientists must "infer history from its results." [1]

Sometimes, we find that the same evidence can be explained in more than one way. When there are competing theories, reasonable people can (and do) disagree about which theory best explains the evidence. Furthermore, in the historical sciences, neither side can directly verify its claims about past events. Fortunately, even

* This title is so long that people usually refer to it simply as *Origin of Species*, or *the Origin*. Actually, the full title is even longer: *"On the Origin of Species by Means of Natural Selection or The Preservation of Favoured Races in the Struggle for Life."*

Figure i:1 Experimental Sciences *Photo: Getty Images*

though we can't directly verify these claims, we can test them. How? First, we gather as much evidence as possible and look at it carefully. Then, we compare the competing theories in light of how well they explain the evidence.[2]

For example, suppose you fall asleep on the couch on a warm weekend afternoon, while watching something on television (a dull movie, let's say). ***On awakening, you step outside and see fact 1 and fact 2:***

From these two facts, or pieces of evidence, what can you conclude? Maybe it rained while you were napping. Maybe the automatic sprinklers came on. With only the data that the driveway and car are wet, both of these explanations are real possibilities.

But suppose you also see the following:

Now what might you conclude? While the sprinkler theory and the rainstorm theory are still possible, these explanations are much less likely in the light of the additional evidence (facts 3 and 4). ***Now, suppose you look a little harder and see this (see top photo, next page):***

With the final piece of data, a new and better explanation for observations 1 through 5 becomes obvious: Someone probably washed the car.

A bucket with soapy water and a sponge sitting behind the car.

If this example seems like everyday common sense to you, it is. It is also an example of making an inference about the past using historical reasoning. You were asleep when whatever happened in the driveway took place. Those events are past; they're history. What remain are clues: signs or pointers (wet places, dry places, weather conditions, buckets, sponges, soapy water, and so on). Your job is to come up with an explanation that makes sense of the clues you see. Starting with this evidence, and using everything else that you know about the world (for example, cars must be washed from time to time, and people often do this in their own driveways), you work backwards in time to what probably happened when you were not around to observe it. The best explanation will be the one that explains more of the evidence than any other. Yes, it is possible that it rained. But only above your driveway? With no clouds in the sky? And would the rainstorm theory explain the bucket of soapy water?

Historical Reasoning and Charles Darwin

Charles Darwin used this same method of historical reasoning when he set out to explain how new life arose on Earth. Although we are now surrounded by living things, no one saw the first plant or animal come into existence. But we have

Figure i:2 Charles Darwin (1809-1882) *Photo Courtesy of the University of Oklahoma History of Science Collections*

plenty of clues. Scientists from many different fields try to piece these clues together to come up with possible explanations. Darwin, himself, looked at many different lines of evidence as he constructed his theory. He considered biogeography (how organisms were distributed over the Earth's surface). He also looked at comparative anatomy (how species resembled each other) and embryology (how organisms develop). Darwin also examined fossils—the mineralized remains of once-living organisms.

Using the clues from each of these areas, Darwin formulated his theory.

Introduction to Darwin's Theory

To understand this book and the issues involved in the discussion, you'll need to know a few key concepts. We'll introduce them here, and examine them in more detail later on. In the *Origin of Species*, Charles Darwin formulated a theory with two main claims.

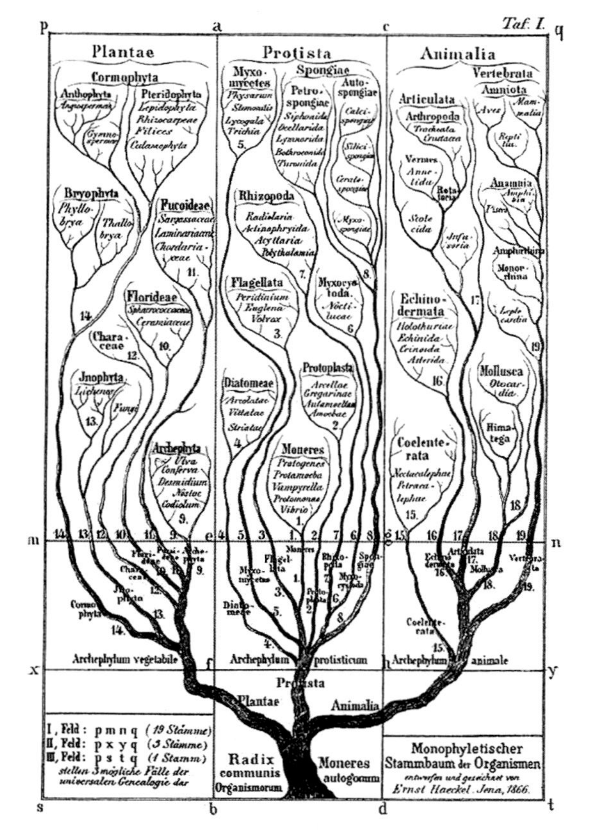

Figure i:3 The tree of life representing life's history, drawn by Ernst Haeckel. *Public Domain.*

The first claim became known as the Theory of Universal Common Descent.[3] This is the idea that every creature on Earth is ultimately descended from a single common ancestor somewhere in the distant past. This theory paints a picture of the history of life on earth—a picture of a great branching tree. Darwin envisioned this "Tree of Life" beginning as a simple one-celled organism that gradually developed and changed over many generations into new and more complex living forms. The first one-celled organism represented the root or trunk of the Tree of Life; the new forms that developed from it were the branches.

The theory's second main claim has to do with the biological process he thought was responsible for this branching pattern. Specifically, Darwin proposed a mechanism that he thought could cause existing living organisms to change, and cause new living forms to arise. Darwin called this mechanism *Natural Selection*, and argued that it had the power to produce fundamentally new forms of life. How could it do that?

Charles Darwin observed that individuals within groups are not exactly the same. Cows from the same herd are not exactly alike. Even puppies in the same litter are not exactly alike. In other words, Darwin observed that organisms *vary* in their traits. Occasionally, these variations between individuals play a huge role in determining which members of the group survive, and which do not.

For example, suppose that sea levels rise dramatically. During high tides, salt water flows into a nearby marsh that previously contained

* Sometimes called "differential survivability and differential reproduction," which is what you should call it when you're trying to impress your parents with how much you're learning.

More Likely To Survive = More Likely To Reproduce

- Variations arise at random.
- Nature "selects" the adaptive (or successful) ones.
- Organisms with the adaptive trait survive and reproduce.
- The offspring are likely to inherit these successful adaptations.
- Inherited adaptations eventually cause populations to change.

only fresh water. Salt is deadly to most plant life, but some plants can tolerate levels of salinity (saltiness) that would kill other organisms. In the new salty environment, the salt-tolerant individuals will probably leave more offspring in the next generation than the non-salt-tolerant (i.e., dead) plants will. The offspring of these salt-tolerant plants are likely to inherit the salt-tolerance trait, which they will likely pass on to their offspring.

Over time, Darwin argued, this process (more likely to survive = more likely to reproduce*) can cause permanent changes in species, and can eventually cause new living forms to arise.

Together, the ideas of Universal Common Descent and natural selection form the core of Darwinian evolutionary theory. They were first spelled out in detail in the *Origin*, and can be found in any biology textbook.

But Darwin's theory itself has evolved a bit since 1859. Darwin (and other biologists of the 19th century) did not understand *how* genetic traits were passed from one generation to the next. In the early decades of the 20th century, biologists learned about the mechanisms of *heredity* (i.e., how traits descend from parents to offspring), and about *mutations* (randomly-arising changes to genetic material, a special kind of variation). The modern evolutionary theory called neo-Darwinism reaffirms the ideas of Universal Common Descent and the creative power of natural selection, and it incorporates this newer knowledge about heredity and mutation that Darwin lacked. Neo-Darwinism is the version of the theory we will examine in this book.

Defining Some Terms

Before we go any further, we have to get a few definitions out on the table. As we look more closely at Darwin's theory, we're going to see that some important terms mean different things to different people. This can be a problem. As the 1990 California Science Framework states, "The process of teaching science requires a precise, unambiguous use of language ... [and] ...Scientists, teachers, and students must communicate the definitions of scientific terms and use them with consistency."[4]

Words must be defined clearly. When someone uses the same word in more than one way, it's called "equivocation."

Let's look at a humorous example of how equivocation can lead to a faulty conclusion:

To be "obtuse" means to be stupid. Some triangles are obtuse. Therefore, some triangles are stupid.

Here's another one:

> *A law can be overturned by the courts or by the legislature. Gravity is a law. Therefore, gravity can be overturned.*

Obviously, both of these conclusions are faulty, and equivocation is to blame in both cases. In each example, an important word was used in two different ways, and the meaning shifted from the first line to the second.

Equivocation has crept into the discussion of *evolution*, as well. Some people use "evolution" to refer to something as simple as small changes in bird beaks. Others use the same word to mean something much more far-reaching. Used one way, the term "evolution" isn't controversial at all; used another way, it's hotly debated. Used equivocally, "evolution" is too imprecise to be useful in a scientific discussion. We're going to take a few moments to define three major uses of "evolution" now, in the hope that it will help us to see where the disagreements really are, and to help us avoid unnecessary arguments later.

Evolution #1: "Change over time."

First, evolution can mean that the life forms we see today are different than the life forms that existed in the distant past. Scientists look in the fossil record to find out what plants and animals existed at different times in Earth's history. What they find, to borrow a phrase, is that the demographics have changed. Most of the plants and animals that are fossilized in recent rock layers are different from the plants and animals fossilized in older rocks. In this definition of evolution, "change over time" refers to changes in the living population as a whole.

Evolution as "change over time" can also refer to minor changes in features of individual species—changes which take place over a short amount of time.[5] Many biologists think this kind of evolution (sometimes called "microevolution") results from a change in the proportion of different variants of a gene within a population. The study of such genetic changes is called *population genetics*.

Evolution #2: "Universal Common Descent."

Some scientists associate the word *evolution* with the idea that all the organisms we see today are descended from a single common ancestor somewhere in the distant past. Evolution in this second sense is an idea we have discussed already: the Theory of Universal Common Descent. As you recall, this theory states that all modern life forms emerged and developed gradually from the first one-celled organism. Since all living things on earth today ultimately have the same ancestor, the history of life is best represented by the picture of a single tree, having many branches, but only one trunk or root.

Evolution #3: "The Creative Power of Natural Selection."

Some people use the term *evolution* to refer to a cause or mechanism of change. When *evolution* is used in this way, it usually refers to the mechanism of natural selection (acting on random variations and mutations). This third use of evolution affirms that the natural selection/mutation mechanism is capable of creating new living forms, and has thus produced the major changes we see in the history of life (as represented by Darwin's Tree of Life).

Keeping Things Clear

Why have we taken so much time with this? We're trying to avoid trouble later. People sometimes get the definitions a little confused, switching from one definition to another in mid-argument. We hope that defining the terms clearly will help prevent this confusion. We will discuss plenty of real disagreements in this book, so we certainly don't need a misunderstood word to create any false disagreements.

The discussion also gets confusing when someone takes evidence for Evolution #1 and tries to make it look like it supports Evolution #2. Conversely, someone may see a problem with Evolution #3, and assume that he or she must reject Evolution #1, as well. This is simply not the case. So, from now on, when you see the word *evolution*, you should ask yourself, "Which of the three definitions is being used?" [6]

Okay, then, which kind(s) of evolution was Darwin talking about? Darwin himself described *The Origin of Species* as "one long argument" for his theory of descent (Evolution #2) with modification (Evolution #3). By "descent," he meant Universal Common Descent. By "modification," he meant the mechanism of natural selection acting on random variations.

Issues in question: Overview

Neo-Darwinism is currently the most widely held view of the history of life. Do all scientists accept all aspects of neo-Darwinism? By no means. What are the sticking points? That depends on whom you ask.

Creative or Conservative?

First, some scientists question whether natural selection can produce the amount of change required by Darwin's Tree of Life scenario. Nearly all biologists agree that natural selection can produce *some* changes in species and that life today is different than it was in the past (Evolution #1). The critical question is, *how much* change can natural selection produce? Some scientists see natural selection as having real—but limited—creative powers.[7] Many of these scientists have begun to doubt whether natural selection can produce fundamentally new forms of life, or major innovations in the anatomical structure of animals (their "body plans"). They see natural selection acting as an *editor,* weeding out harmful variations in body design, while conserving (keeping) helpful variations.

In contrast, neo-Darwinists also see natural selection acting as a *writer,* capable of true creativity and innovation. Neo-Darwinian biologist Francisco Ayala, for example, affirms that the "creative duet" of mutation and natural selection can produce the "organization of living beings."[8] Zoologist Ernst Mayr writes that natural selection is a "positive, constructive force," and adds "one can go even further and call natural selection a creative force."[9] It is important to remember that when neo-Darwinists refer to the creative power of natural selection, they are *not* claiming that natural selection is "trying" to create anything in particular. Instead, they mean that the undirected mechanism of natural selection acting on random variations can produce the fundamentally new structures and forms of life that arise during the history of life. They affirm what zoologist Richard Dawkins calls "the power of natural selection to put together good designs." [10]

Tree or Orchard?

Second, neo-Darwinists contend that "a single Tree of Life containing multiple branches" is the most accurate picture of the history of life. Other scientists doubt that all organisms have descended from one—and *only* one—common ancestor. They say that the evidence does indeed show some branching taking place *within* larger

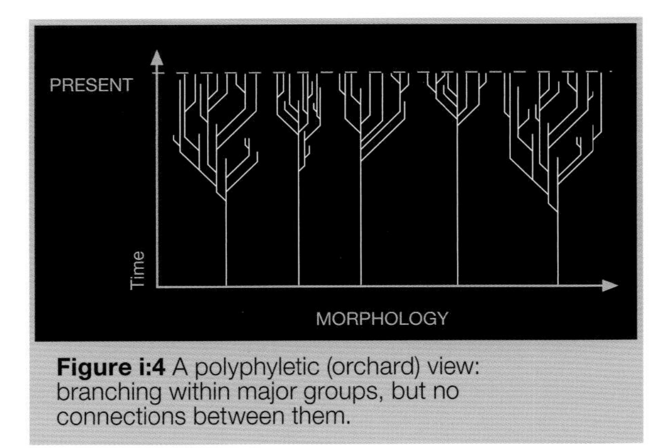

Figure i:4 A polyphyletic (orchard) view: branching within major groups, but no connections between them.

groups of organisms, but not *between* the larger groups. According to these scientists, the history of life should not be represented as a single tree, but as a series of parallel lines representing an orchard of distinct trees. In the orchard view, each of the trees has a separate beginning.

Which picture best illustrates the history of life? Scientists have different views, and these competing views have names. Scientists who think that the history of life is best represented by a single branching tree have what is called a *monophyletic* view ("mono" means one or single). Scientists who have a *polyphyletic* view ("poly" means many) think that the history of life looks more like an orchard of separate trees.[11]

As part of this tree discussion, we have to make an important distinction between the terms *common descent* and *Universal Common Descent*. You may think these terms mean the same thing. They don't. As we've just seen, it's possible to think that some organisms share a common ancestor without thinking that *all* organisms are descended from a *single* common ancestor.

To help you keep this important distinction in mind, we're going to use the term *Common Descent* (with capital letters) to refer to Darwin's idea that all organisms share a single common ancestor—the theory of Univeral Common Descent or the monophyletic view of the history of life.

When we use the term *common descent* (no capitals), we're referring to limited common descent—the view that separate groups of organisms have common ancestors.

The Stage is Set

Do all living things, past and present, share a common ancestor? Can natural selection produce fundamentally new organisms from pre-existing ones? The rest of this book will examine the claims of scientists who disagree about the answers to these questions. Evidence that seems convincing to one scientist can seem inadequate to another, while yet another scientist may think the evidence points in a different direction.

Looking at the evidence and comparing the competing explanations will provide the most reliable path to discovering which theory, if any, gives the best account of the evidence at hand. Making that comparison is your job. We're asking you to be part scientist, part detective, and part juror. Lots of people will try to influence a jury's decision about which side is right, and why. Keep your wits about you. In science, it is ultimately the evidence—and *all* of the evidence—that should tell us which theory offers the best explanation. This book will help you explore that evidence. Scientists working to find better theories often re-examine and refine their original ideas in the light of new data. This is certainly true in the debate about the history of life. We are always learning more about the world, and this often involves new discoveries on the cutting edge of research. We hope this book will stimulate your interest in these questions as you weigh the competing arguments.

Endnotes

1 Stephen Jay Gould, "Evolution and the triumph of homology: Or, why history matters," *American Scientist* 74 (1986):61.

2 C. Cleland, "Historical science, experimental science, and the scientific method," *Geology* 29 (2001):987-990.

C. Cleland. "Methodological and epistemic differences between historical science and experimental science," *Philosophy of Science* 69 (2002):474-496.

3 Darwin envisioned this process of "descent with modification" continuing until it had produced all the organisms we see today. In a famous passage at the end of *the Origin*, Darwin argued that, "all the organic beings which have ever lived on this earth have descended from some one primordial form." This theory is called *Universal Common Descent* because it claims that every organism on Earth is connected to the same tree of life, rooted in the same common ancestor. (The "universe" in this case refers only to life on our planet, not the entire physical universe.) Charles Darwin, *On the Origin of Species* (Cambridge, Mass: Harvard University Press, 1964 [Facsimile of the First Edition, 1859]):484. Elsewhere in the *Origin*, Darwin allowed for the possibility that life might have arisen from one *or from a few* original life forms.

4 *California Science Framework*, 1990:14 & 17.

5 Scientists define "species" in many different ways. (There are 25 different definitions at last count.) One commonly used definition is "populations of interbreeding organisms that are reproductively isolated from other such populations."

6 Richard J. Bird, *Chaos and Life: Complexity and Order in Evolution and Thought* (New York: Columbia University Press, 2003).

7 Marc Kirschner and John Gerhart, *The Plausibility of Life: Resolving Darwin's Dilemma* (New Haven: Yale University Press, 2005).

Wallace Arthur, *Biased Embryos and Evolution* (Cambridge: Cambridge University Press, 2004).

Wallace Arthur, *The Origin of Animal Body Plans: A Study in Evolutionary Developmental Biology* (Cambridge: Cambridge University Press, 1997).

Scott F. Gilbert, John M. Opitz, and Rudolf A. Raff, "Resynthesizing evolutionary and developmental biology," *Developmental Biology* 173 (1996):357-372.

Brian Goodwin, *How the Leopard Changed Its Spots: The Evolution of Complexity* (New York: Charles Scribner's Sons, 1994).

George L. Gabor Miklos and Bernard John, "From genome to phenotype," in *Rates of Evolution*, K. S. W. Campbell and M. F. Day, eds. (London: Allen and Unwin, 1987).

George L. Gabor Miklos, "Emergence of organizational complexities during metazoan evolution: perspectives from molecular biology, palaeontology and neo-Darwinism," *Memoirs of the Association of Australasian Palaeontologists* 15 (1993):7-41.

G. Webster and B. C. Goodwin, "The origin of species: a structuralist approach," *Journal of Social and Biological Structures* 5 (1982):15-47.

8 Francisco Ayala, *Creative Evolution!?* eds. J.H. Campbell and J.W. Schopf (Sundberg, Mass: Jones and Bartlett, 1994):4-5.

9 Ernst Mayr, Introduction to *On the Origin of Species* by Charles Darwin, (Cambridge, Mass: Harvard University Press, 1964 [Facsimile of the First Edition, 1859]): xvii; and Ernst Mayr, "Accident or Design, The Paradox of Evolution," in G. Leeper, ed., *The Evolution of Living Organisms* (Cambridge: Cambridge University Press, 1962):1-8.

10 Richard Dawkins, *The Blind Watchmaker* (New York: W.W. Norton, 1987):95.

11 Scientists who support a polyphyletic view differ on how many trees one should expect to find in the "orchard" of life. Some, such as microbiologist Carl Woese of the University of Illinois, argue that "life on Earth is descended not from one, but from three distinctly different cell types," ("On the evolution of cells," *Proceedings of the National Academy of Sciences* 99 [2002]:8742-77; 8746). Others, including Malcolm Gordon of UCLA and Christian Schwabe of the Medical University of South Carolina, think there might be a greater number of separate trees.

Malcolm S. Gordon, "The concept of monophyly: a speculative essay," *Biology and Philosophy* 14 (1999):331-348.

Leo S. Berg, *Nomogenesis or Evolution Determined By Law* (Cambridge: The M.I.T. Press, 1969).

Christian Schwabe and Gregory W. Warr, "A polyphyletic view of evolution: the genetic potential hypothesis," *Perspectives in Biology and Medicine* 27 (1984):465-485.

Christian Schwabe, "Genomic potential hypothesis of evolution: a concept of biogenesis in habitable spaces of the universe," *The Anatomical Record* 268 (2002):171-179.

Christian Schwabe, "Theoretical limitations of molecular phylogenetics and the evolution of relaxins," *Comparative Biochemistry and Physiology* 107B (1994):167-177.

Michael Syvanen, "Recent emergence of the modern genetic code: a proposal," *Trends in Genetics* 18 (2002):245-248.

Carl R. Woese, "A new biology for a new century," *Microbiology and Molecular Biology Reviews* 68 (2004):173-186.

G. Webster and B.C. Goodwin, "The origin of species: a structuralist approach," *Journal of Social and Biological Structures* 5 (1982):15-47.

D'arcy Wentworth Thompson, *On Growth and Form* (New York: Dover Publications, Inc. 1992).

Universal Common Descent
Arguments for and Against

As you recall, both the original and the contemporary versions of Darwinian theory make two main claims. One claim is that all organisms descended from a common ancestor. So, our first task will be to examine the five main lines of evidence that Darwin used, or that modern neo-Darwinists use, to support this part of evolutionary theory. For each line of evidence, we shall see how some scientists interpret this same evidence differently. In some cases, we will look at additional evidence that scientists see as strongly challenging this part of neo-Darwinian theory.

Later in this book, we'll look at neo-Darwinism's other main claim: that the mechanism of Natural Selection and random mutation has the creative power to produce fundamentally new organisms from earlier, simpler forms of life. Then, we will again examine the scientific evidence and counterarguments that have led some scientists to challenge this part of modern Darwinian theory.

The purpose of this book is to introduce you to both the case *for* and the case *against* the major aspects of neo-Darwinian evolutionary theory. You may be surprised to learn that the scientific literature contains evidence-based arguments on both sides of each issue—some supporting, and some challenging each key aspect of the theory.

Let's start by examining the five main arguments for and against the Theory of Universal Common Descent.* We will begin by examining what the fossil evidence has to say. ■

*The arguments for Common Descent are: fossil progression, anatomical homology, molecular homology, embryological similarity and biogeographical distribution.

Fossil Succession

Fossil Succession: Case For

Have you ever wondered what life was like a long, long time ago? You're not alone. Many scientists have spent their entire careers studying the history of life on earth. Some of them do this by examining fossils.

Fossils are the mineralized remains or impressions of organisms that lived in the past. Both plants and animals can be fossilized. Sometimes the fossilized organism was buried in sediment. You could think of it like something getting buried in quicksand. When the sediment fully hardens (or "lithifies"), it creates a permanent impression of the organism in the rock. Fossils can be found in rocks all over the Earth, and sometimes we find a whole lot of them in the same place.

Darwin thought that fossils told us a lot about the history of life on Earth. Specifically, he thought the sequences of fossil forms found in sedimentary rocks supported one of his central ideas: Common Descent. Here's why.

Darwin pointed out what geologists already knew: that if you dig a pit deep enough, you will notice that the rocks in the walls of the pit are arranged in layers. Scientists call these layers strata *(see Figure 1:1)*. Generally speaking, rocks in the deeper strata are older than rocks in the shallower strata.*

Darwin also knew that we sometimes find different fossils in one layer than we find in the next layer. Some strata contain no fossils at all. Other strata contain fossils such as trilobites and mollusks *(see Figure 1:2)*, but not birds or mammals. In another, one might find fossil horses or pigs—or primates (the mammalian order to which humans belong).

It seemed obvious to Darwin that life had changed over the course of geological history. He also thought these changes fell into some recognizable patterns.

What patterns did he see? As he looked at fossil records from many different places, all the way up and down the strata, he noticed a pattern of appearances and disappearances. For example, we first find trilobites in the Cambrian strata. We continue to find trilobite fossils as we move upward through the strata, in rock

* Sometimes, geological processes like thrust faulting and folding can push older rocks above younger rocks. Most of the time, however, deeper rocks are older.

Figure 1:1 (left) An example of strata *Photo: iStock.* (right) The appearance of new forms over time in the fossil record. Time moving from bottom to top, earlier to more recent.

layers covering a period of about 300 million years. Then, the trilobites disappear forever—or become *extinct*—in the Permian strata, higher up in the rock record. Later on, the first mammals appear—in the Triassic period, long after the trilobites are gone. Darwin called this trend of fossil appearances and disappearances "geological succession." Modern scientists call it "fossil succession."

But this was not the only trend Darwin saw. The fossil record seemed to show a trend from simple to complex. In other words, it seemed clear to Darwin that as he looked higher and higher in the fossil record, he found more and more complicated creatures fossilized there.

And he noticed one more important feature about how life had changed over time.

A Branching Pattern

To see what Darwin saw, let's imagine what a graph of fossil progression might look like. The present time is at the top of the graph. The formation of the Earth is at the bottom. Each time a new type of animal comes into the fossil record, we'll put a dot on the graph. The earlier the animal comes into the fossil record, the lower the dot will be on the graph.

Draw a line upward from the "first appearance" dot. Stop drawing the line at the time when we stop finding fossils of that animal. Place another {cont. on page 19}

Figure 1:2 Geological Timeline *(source: International Commision on Stratigraphy)*

ERAS	PERIODS	ALTERNATE PERIODS	EPOCHS (North America)	DURATION (in millions of years)	
Cenozoic Era	Neogene Period	Quarternary Period	Holocene Epoch	23-24 M.Y.	1.8 M.Y.
			Pleistocene Epoch		
		Tertiary Period	Pliocene Epoch		64 M.Y.
			Miocene Epoch		
	Paleogene Period		Oligocene Epoch	42 M.Y.	
			Eocene Epoch		
			Paleocene Epoch		
			65.8 M.Y. Ago		
Mesozoic Era	Cretaceous Period			79.7 M.Y.	
			145.5 M.Y. Ago		
	Jurassic Period			54.1 M.Y.	
			199.6 M.Y. Ago		
	Triassic Period			51.4 M.Y.	
			251 M.Y. Ago		
Paleozoic Era	Permian Period			48 M.Y.	
			299 M.Y. Ago		
	Carboniferous Period			60.2 M.Y.	
			359.2 M.Y. Ago		
	Devonian Period			56.8 M.Y.	
			416 M.Y. Ago		
	Silurian Period			27.7 M.Y.	
			443.7 M.Y. Ago		
	Ordovician Period			44.6 M.Y.	
			488.3 M.Y. Ago		
	Cambrian Period			53.7 M.Y.	
			542 M.Y. Ago		
Neoproterozoic Era (Precambrian Time)	Ediacaran Period			88 M.Y.	
			630 M.Y. Ago		
	Other Precambrian Periods			Approx. 4,000 M.Y.	

dot at that point. The longer the line, the longer the species lasted. The farther apart one line is from another, the greater the difference in the anatomy of the two types of animals. We'll repeat this process for each animal species we know about. We should now have a graph full of lines where each line represents a new form of life.

Now, imagine what would happen if you connected the dots, connecting a dot from one line to the starting dot of another line. Does the new drawing suggest a branching tree to you? It did to Darwin. His famous tree analogy was Darwin's way of interpreting (or making sense of) the fossil data. But what sense *did* he make of it?

The first major lesson Darwin drew from this diagram was that younger fossil forms arose from older ones. Why do the first mammal fossils appear so long after the first reptiles? In Darwin's

view, it's because mammals have reptiles as their ancestors. The mammal group branched out from the reptile branch at some point in the past and developed into its own branch. Reptiles and mammals, in turn, must have branched out from another vertebrate group—the distant ancestor they last had in common.

The second major lesson Darwin drew was that every creature on Earth must ultimately be linked to a single common ancestor in the distant past: the root or trunk of the Tree of Life. Darwin thought of life beginning with one or very few original forms, while most neo-Darwinists maintain that there is only one original form, called the Last Universal Common Ancestor (or LUCA).[1]

Fossil Succession Suggests Darwin's Tree of Life

But let's back up a moment. Why did Darwin think we should "connect the dots" on our graph? It's true that we see one fossil line here, and another one over there. But for Darwin's branching Tree of Life to be an accurate picture of life's history, there must have been some "in-

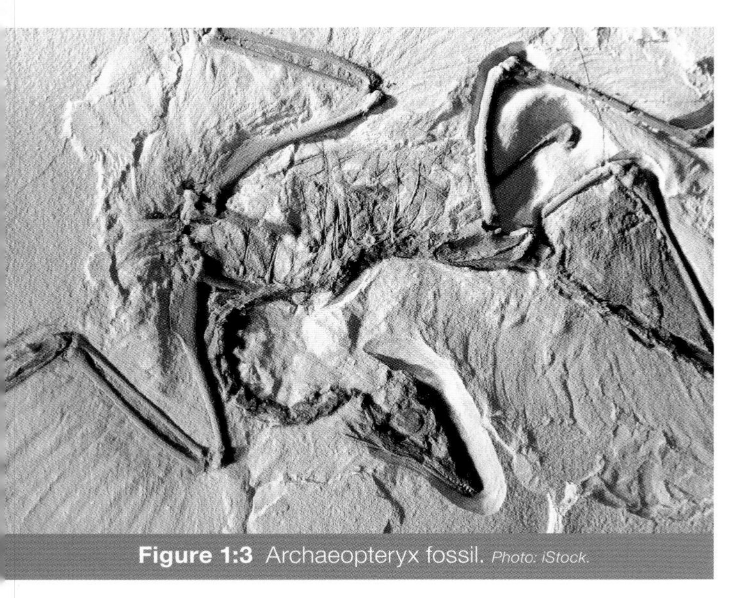

Figure 1:3 Archaeopteryx fossil. *Photo: iStock.*

between" forms that existed between one branch and another. Evolutionary biologists call these in-between forms "intermediate" or "transitional" forms. They argue that such forms have, in fact, been found.

For instance, they point to fossils called "mammal-like reptiles," which appear in Permian and Triassic strata (200 – 300 million years ago). Mammal-like reptiles are extinct groups that appear to have mostly reptilian traits, mixed in with some mammalian features. Or consider another example. Recently, some scientists think they have discovered a transitional fossil sequence connecting land-dwelling mammals to whales.[2]

Both of these transitional sequences were discovered after Darwin's time, but he was alive when another possible transitional form, *Archaeopteryx*, was discovered in a limestone quarry in Germany. *Archaeopteryx* was a bird with a toothed jaw like a reptile, but true feathers like a modern bird. Darwin celebrated this discovery because it was the very sort of fossil his theory predicted we would find. As historians Adrian Desmond and James Moore have noted, "Darwin… reveled in the splendid 'bird-creature with its long tail and fingers.'"[3]

Paleontologists have identified many gaps that remain to be filled in the fossil record.[4] Typical evolutionary tree diagrams represent these gaps as dotted lines *(see Figure 1:5)*—pieces of the tree that we have yet to find. But advocates of Common

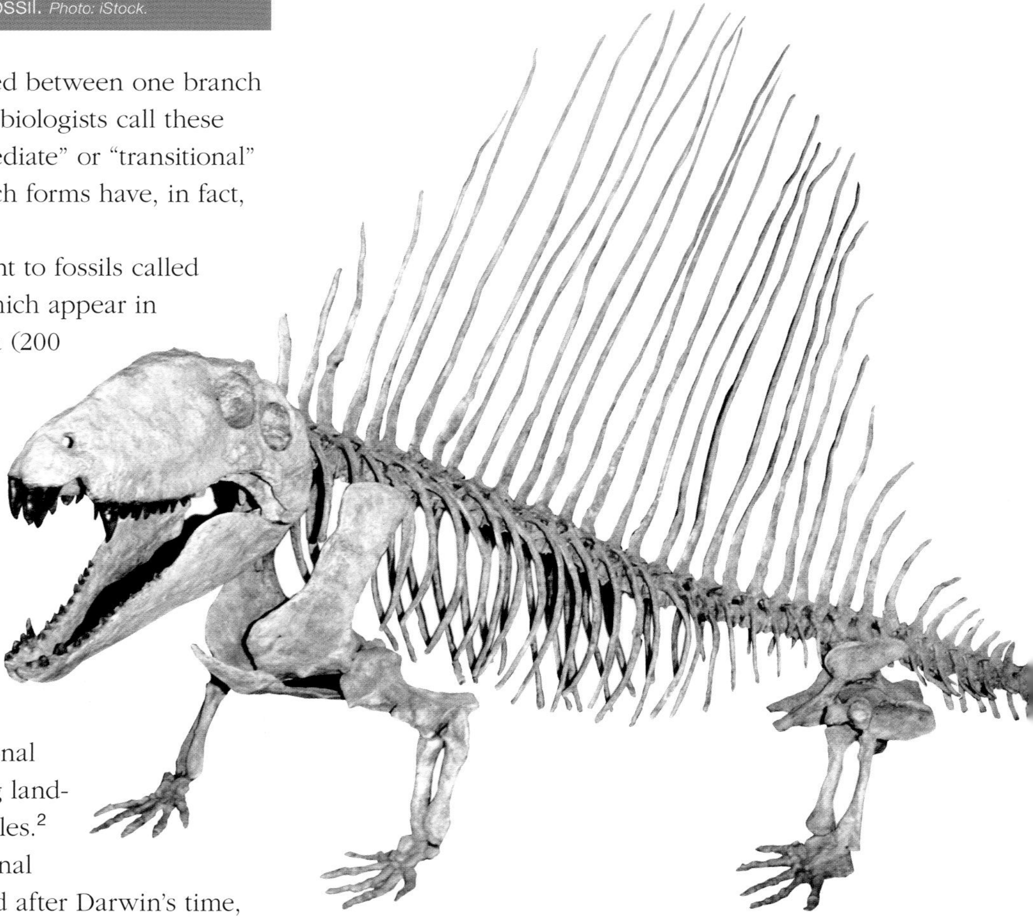

Figure 1:4 Dimetrodon: a mammal-like reptile.
Photo by Ken Lucas/Visuals Unlimited

Descent contend that the gaps that have been filled make it reasonable to assume that we'll eventually find the rest. Therefore, they say, the "single branching tree" model best represents the history of life and best explains the patterns we observe in the fossil record.

Advocates of Common Descent also point out that fossils often appear in the order predicted by other kinds of evidence. By analyzing the similiarities and differences of organisms, evolutionary biologists can predict the order in which organisms should appear in the fossil record. For example, based on *Archaeopteryx's* combination of dinosaur and bird-like features, evolutionary biologists expect to find *Archaeopteryx* coming in the fossil record after dinosaurs and before birds. A recent fossil find suggests that two-footed carnivorous dinosaurs (those called theropods) may have arisen before *Archaeopteryx.* This discovery has given evolutionary biologists further confidence that *Archaeopteryx* is a transitional form linking dinosaurs to birds.

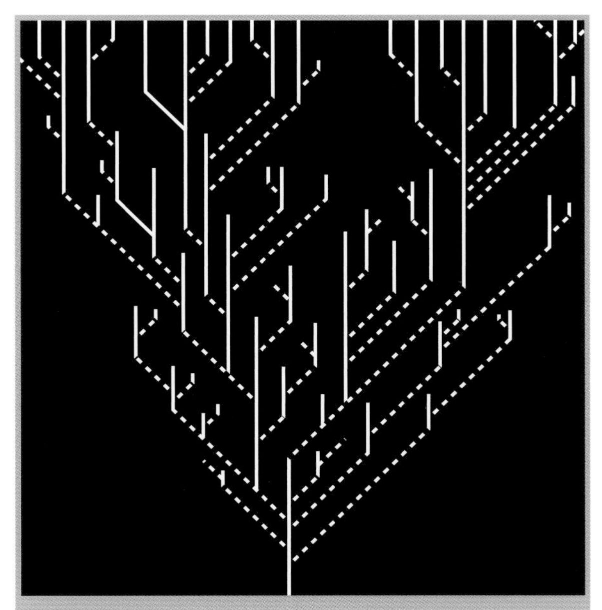

Figure 1:5 Tree of Life, with dotted lines showing gaps still to be filled.

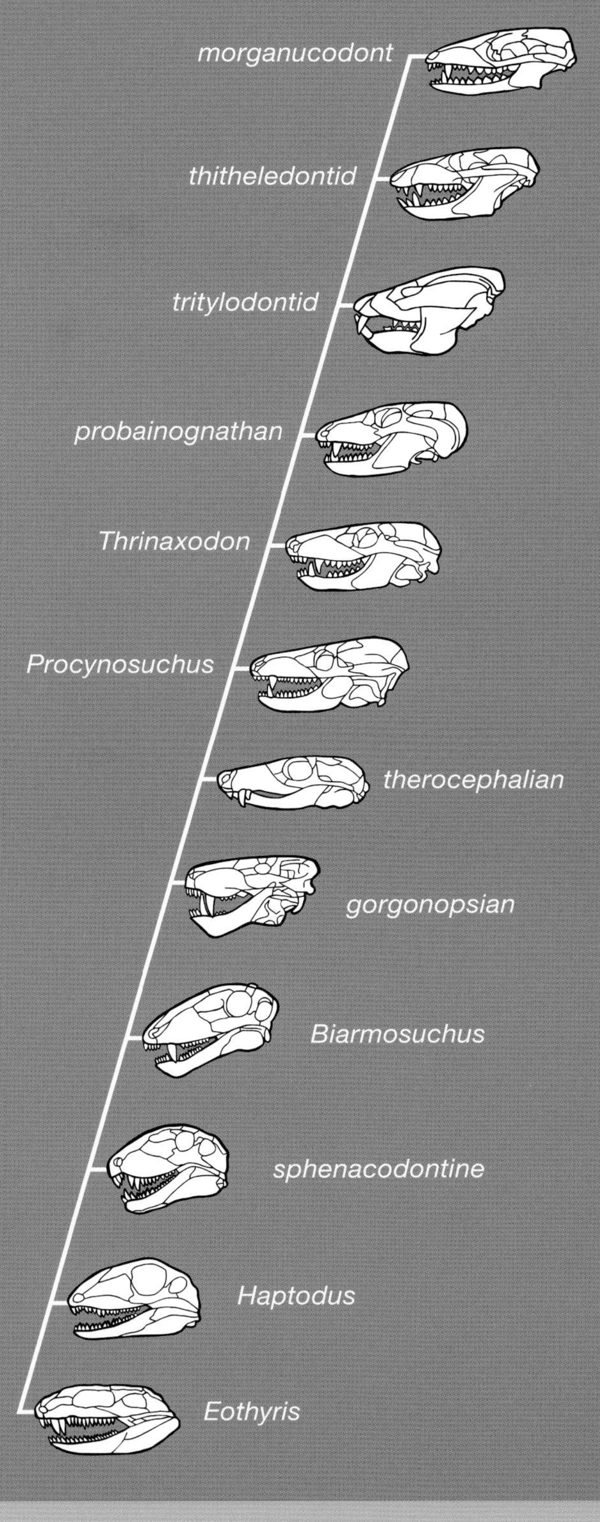

Figure 1:6 Sequence of Mammal-like reptiles as typically presented in textbooks.
From T.S. Kemp, The Origin & Evolution of Mammals *(2005): 89.*

Fossil Succession: Reply

Most critics of the fossil succession argument agree that the fossil record shows change over time. They also accept that more recent animal forms are generally—although not always—more complex than the forms found deeper in the fossil record. Nevertheless, they contend that the overall pattern of fossil evidence contradicts the neo-Darwinian picture of the history of life in two important respects.*

First, paleontologists have discovered that new animal forms almost always appear abruptly—not gradually—in the fossil record, without any obvious connections to the animals that came before.[5]

Abrupt Appearance

For example, about 530 million years ago, more than half of the major animal groups (called phyla) appear suddenly in the fossil record.[6] In biological classification the term "phylum" (plural: phyla) designates a large group of animals that share a unique or distinctive body plan (see sidebar).

Because this sudden and dramatic appearance occurred in the Cambrian period, paleontologists now refer to this event as "the Cambrian explosion." Many paleontologists now estimate the Cambrian explosion took place over a period of 10 million years or less.[7] This seems like a long time to us, but it's a blink of an eye in Earth-time. If Earth's whole history were a timeline the length of an American football field, the Cambrian explosion time would take up just 4 inches of the football field's total length. It is only two-tenths of one percent of geological history. More importantly, many evolutionary biologists doubt that this is enough time for the slow, gradual Darwinian processes to produce the amount of change that arises in the Cambrian explosion.[8]

For this reason, many scientists think this geologically sudden appearance of many new life forms contradicts Darwin's prediction that new forms would emerge gradually over vast spans of geological time.

What's true of the phyla (the highest animal classification) is also true of the middle and of many lower classifications. These also appear suddenly. For example, in the Paleocene epoch (65-55 million years ago), 15 new mammalian orders suddenly enter the fossil record.** Scientists call this series of events the "mammalian {cont. on page 24}

*Trilobites, for example, are among the earliest animals, appearing during the Cambrian explosion. Yet, they were quite complex, having dozens of cell types, and compound eyes using complicated optics.

Biological Classification

Scientists divide life into three basic groups or *domains*: Bacteria, Archaea, and Eukarya. Each of these has a fundamentally different cell structure. Within Domain Eukarya—creatures whose cells have nuclei, as we do—we find the Kingdom Animalia. Within Kingdom Animalia, we find radically different body designs. For example, one type of body plan has an internal skeleton and a backbone, another plan has a jellyfish-like body, while another body plan has a segmented body. Scientists classify each unique animal body plan as a separate phylum within the animal kingdom. The bumblebee belongs to the Phylum Arthropoda, as do all other insects. The polar bear, along with all other mammals, is classified in the Phylum Chordata.

Within each phylum, there are other distinguishing features that allow biologists to further separate animals into other, smaller categories. These include classes, then orders, and so on, all the way down the line to species.

The higher levels of biological classification—like kingdom, phylum, and class—describe life's major categories. Lower levels of biological classification—like family, genus, and species—designate smaller degrees of differences within the major categories. ■

	Bumblebee	Polar Bear
Domain	Eukarya	Eukarya
Kingdom	Animalia	Animalia
Phylum	Arthropoda	Chordata
Class	Insecta	Mammalia
Order	Hymenoptera	Carnivora
Family	Apidae	Ursidae
Genus	*Bombus*	*Ursus*
Species	*terricola*	*maritimus*

** How big are the distinctions between orders? Think of this: Carnivora (e.g., bears), Chiroptera (e.g., bats), and Perissodactyla (e.g., horses) are all different orders within the mammal class. To have 15 mammalian orders appear suddenly within the same narrow window of geologic time is a really big deal.

Coming Out With Their Shell

Turtles are another fascinating example of a group of animals that appears abruptly in the fossil record. The order Chelonia, to which turtles and tortoises belong, appears suddenly in the late Triassic, around 200 million years ago. The very first time turtles appear, their body plan is already fully developed, and they appear in the fossil record without intermediates. Furthermore, turtle and tortoise shells contain more than 50 bony "scutes" that appear in no other vertebrate order, nor anywhere else in the fossil record. What's more, the turtle scapula is positioned underneath its ribs and scutes, unlike any living or fossilized vertebrate. Scott Gilbert, an evolutionary biologist who works on this puzzle, says that "the turtle shell represents a classic evolutionary problem: the appearance of a major structural adaptation." According to Gilbert, this problem is made even more difficult by "the 'instantaneous' appearance of this evolutionary novelty."* Because "the distinctive morphology of the turtle appears to have arisen suddenly," Gilbert and his colleagues argue that evolution needs "to explain the rapid origin of the turtle carapace [shell]." They are studying turtle embryology to investigate how this might have happened.** ■

*Scott Gilbert, "Morphogenesis of the Turtle Shell: the Development of a Novel Structure in Tetrapod Evolution," *Evolution & Development* 3 (March-April 2001):56.

**Judith Cebra-Thomas, Fraser Tan, Seeta Sistla, Eileen Estes, Gunes Bender, Christine Kim, Paul Riccio, and Scott F. Gilbert, "How the Turtle Forms Its Shell: A Paracrine Hypothesis of Carapace Formation," *Journal of Experimental Zoology* (Mol Dev Evol) 304B [2005]:1-12.

radiation." Not only do new mammalian orders appear suddenly, but when they appear, they are already separated into their distinctive forms. For example, during the Eocene epoch (just after the Paleocene), the first fossil bat appears suddenly in the fossil record. When it does, it is unquestionably a bat, capable of true flight.[9] Yet, we find nothing resembling a bat in the earlier rocks.

Critics of the fossil succession argument point out that what is true of animals is also true of plants. For example, flowering plants appear suddenly in the early Cretaceous period, 145-125 million years ago. This rapid appearance is sometimes called the angiosperm big bloom. "The origin of the angiosperms remains unclear," writes one team of researchers. "Angiosperms appear rather suddenly in the fossil record…with no obvious ancestors for a period of 80-90 million years before their appearance."[10] This contradiction was so perplexing that Darwin himself referred to it as "an abominable mystery."[11]

As a result, critics say the pattern of fossil appearance does not support Darwin's picture of a gradually branching tree. {cont. on page 26}

Fossil to Modern Comparisons

The fossil record provides many examples of living organisms that have remained stable in their form and structure over many millions of years—sometimes over *hundreds* of millions of years. These photos show several examples of such stability over time. The top two sets of pictures compare the internal and external structure of a chambered nautilus shells. The photos on the left show pictures of fossilized nautilus shells from the Devonian period. The photos on the right show modern nautilus shells, virtually unchanged after 400 million years of geo-logic time. The third set of pictures compares fossilized Ginkgo leaves with modern Ginkgo leaves, showing no change in structure in 135 million years. The final set of pictures is perhaps one of the most dramatic examples. The picture on the left shows a fossilized comb jelly from the Cambrian period, 530 million years ago. The picture on the right shows a modern, living comb jelly at home in the water. The form of the two organisms is identical. ■

Fossil Nautilus Shell Photo: Alex Kerstitch/Visuals Unlimited; Modern Nautilus Shell Photo: iStock Images; Fossil Shell Cut Away: iStock Images; Modern Shell Cut Away: iStock Images; Fossil Ginko Leaf Photo: Brian Gage; Modern Ginkgo Leaves Photo: iStock Photo; Fossil Comb Jelly Photo: Chen, et al.; Modern Comb Jelly Photo: Tom Stack Photo.

Instead, it gives the distinct impression of a series of independent beginnings. For this reason, Darwin himself said that the pattern of abrupt appearance (his own term), "may be truly urged as a valid argument" against his theory of Common Descent.[12]

The Stability of Biological Forms

Critics of fossil succession point to a second feature of the fossil record that challenges the Darwinian picture of the history of life. Recent fossil studies reveal that most animal forms remain relatively stable throughout their time on earth. Paleontologists call this stability of form *stasis* when it occurs at the species level. But stability also characterizes the body designs of the organisms representing the higher categories of life—the orders, classes and phyla.[13] "Instead of finding the gradual unfolding of life," writes paleontologist David Raup of the University of Chicago, "what geologists of Darwin's time, and geologists of the present day actually find is a highly uneven or jerky record; that is, species appear in the sequence very suddenly, show little or no change during their existence in the record, then abruptly go out of the record."[14]

According to Joel Cracraft, a systematist at the American Museum of Natural History, *On the Origin of Species* contains no geological evidence to support Darwin's contention that evolution was a slow, gradual process. "The fossil record, seemingly so important for anyone advocating evolutionary modification through time, was not very kind to Darwin's cause. As a result, he ignored it; the fossil record certainly did not make him alter his theorizations or expectations. In fact, what he often saw was stasis."[15]

The sudden appearance of major new forms of life, and the stability of these forms over time, have led some scientists to doubt that the fossil record supports the case for Common Descent.[16]

But what about transitional intermediates like *Archaeopteryx* and the mammal-like reptiles?

If the fossil record is as disconnected as the critics say, then why does it show evidence of transitional forms, such as *Archaeopteryx* and the mammal-like reptiles? Doesn't this evidence show that Darwin's picture of the history of life was right after all?

No, say the critics. Here's why. Imagine that you are a teacher's assistant, grading a math test with 100 questions. One student hands in a paper with three right answers—and all the rest wrong. The student draws your attention to the three correct answers, and says, "This proves that I understand the concept. It was just a fluke that I missed the others." How would you respond?

Chances are, you would point to the 97 red check marks and say, "*A fluke?* What test are *you* looking at? I'm looking at 97 wrong answers. If you want to talk about flukes, check out the three right answers."

In much the same way, critics point out that discontinuity (abrupt appearance, followed by stasis) is the prevailing pattern of the fossil record.[17] The transitional forms are the rare exceptions. Even advocates of the Darwinian account acknowledge that the fossil record displays far fewer transitionals than predicted by the theory of Common Descent. For example, in 1982, Oxford University paleontologist Thomas Kemp noted that, "Indeed, [the mammal-like reptile series] is the only such major transition in the animal kingdom that is anything like well-documented by an actual fossil record."[18]

Though a possible whale-to-mammal transitional sequence has recently been unearthed, critics maintain that transitional sequences are rare, at best. For this reason, critics argue that Darwin's theory has failed an important test. Just as students are tested by exams, theories are tested by how well they match the evidence. In the overwhelming majority of cases, Common Descent does not match the evidence of the fossil record. A student who gets a correct answer only once in a while does not deserve a passing grade. In the same way, critics say that a scientific theory that only rarely matches the evidence fails the test of experience.

Reasons for Suspicion

There's another aspect to this problem. Scientific critics of the Fossil Succession argument doubt that the alleged transitional sequences in the fossil record are really what they appear to be in the textbooks.

Sequence Problems

Some critics are unpersuaded for statistical reasons by the few transitional sequences that have been found. Given the millions of different fossil forms in the fossil record, critics argue that we would expect to find, if only by pure chance, at least a few fossil forms that could be arranged in plausible evolutionary sequences. To understand what they mean, imagine that a representative of every organism that has ever lived on earth was randomly pasted to an enormous wall representing the geologic column. Most of the fossils would bear no relationship to the other fossils stuck closest to them on the wall. Nevertheless, by chance a few of them might end up next to forms that do have some resemblance. These forms might then appear to be related as ancestors and descendents, even if they were not. Is it possible that the mammal-like reptile sequence is a statistical anomaly rather than a legitimate sequence of ancestors and descendents?

Another problem is that fossils don't always appear in the order they're predicted to. Evolutionary biologists analyze the features of organisms looking for similarity. From these analyses, they generate hypothetical branching-tree diagrams called cladograms. Scientists use these diagrams to predict which organisms should appear early in the fossil record, and which should appear later.

These predictions often do not match the actual appearance of animal fossils in the fossil record, (though they are remarkably consistent for plants). Many "older" groups of animals (as depicted in cladograms) appear above, not below, the supposedly "younger" ones in the fossil record. The primate fossil record "poorly reflects" the predicted evolutionary sequence, say Norrell and Novacek. "Groups thought to have branched off early in primate history appear late in the record or have no fossil record."[19] The problem is not as serious with the mammal-like reptiles. However, five "intermediate" forms that cladograms predict should have arrived neatly in sequence over a long period of time actually appear suddenly at the same time in the fossil record.[20]

Size Problems

Some textbooks alter the scale of pictures showing the order of appearance of groups such as the mammal-like reptiles. This makes the features appear closer in size *{cont. on page 29}*

“Out of the Water and On to the Land?”

Scientists have long thought that amphibians were a transitional form between aquatic and land-dwelling life forms. Why? Because amphibians can live in both the water and on land. Yet, the fossil record has revealed at least two problems with this idea.

First, biologist Malcolm Gordon and paleontologist Everett Olson point out that land-dwelling amphibians, themselves, appear suddenly in the fossil record. They first show up in the late Devonian period, with no apparent connection to earlier life forms. Gordon and Olson point out that the earliest amphibian fossils unmistakably show them as four-footed creatures. They go on to say that we have no fossil evidence that relates directly to the vertebrate transitions to land, and no fossil evidence of transitional stages in earlier rocks.* In the picture below, all of the creatures above the hypothetical transition line are already true tetrapods.

Second, the earliest fossil evidence of amphibians comes from sites in Greenland, South America, Russia, and Australia. During the late Devonian period, these sites were “separated from one another by thousands of kilometers of open ocean and land.”** Since the first amphibian fossils appear at the same time, yet are separated by such large distances, it would appear that the *same* transition was taking place *simultaneously* in multiple locations—a conclusion many scientists find improbable.

The large distances and the absence of transitionals have led Gordon and Olson to think these tetrapods arose multiple times, independently of one another.*** More recently, paleontologists have found fossils that seemed to show a connection between fish and tetrapods—in particular, in the structure of the front fins of some bony fish and in the forelimbs of an early tetrapod.**** Will these finds establish the common ancestry of all tetrapods which Gordon and Olson have questioned? Stay tuned. ■

---

*“No fossils are know[n] that relate directly to the vertebrate transitions to land.” Instead, these creatures were definitely tetrapods “…by the time that the first fossil amphibians are recorded, and no tangible evidence of transitional stages has yet been found in earlier rocks.” Malcolm S. Gordon and Everett C. Olson, *Invasions of the Land: The Transitions of Organisms from Aquatic to Terrestrial Life* (Columbia University Press: New York, 1995):128-133, 262-264.

** Malcolm S. Gordon and Everett C. Olson, *op. cit.*

***The creatures were already “structurally diverse, fairly specialized, and phylogenetically well differentiated from one another…. The geographic distribution and morphological diversity of the fragmentary remains has posed problems and has led to controversy as to whether amphibians are monophyletic or polyphyletic.” Malcolm S. Gordon and Everett C. Olson, *op. cit.*

John A. Long and Malcolm S. Gordon, “The greatest step in vertebrate history: A paleobiological review of the fish-tetrapod transition,” *Physiological and Biochemical Zoology* 77 (2004):700-719.

**** Jennifer A. Clack, “From Fins To Fingers,” *Science* 304 (April 2, 2004):57-58.

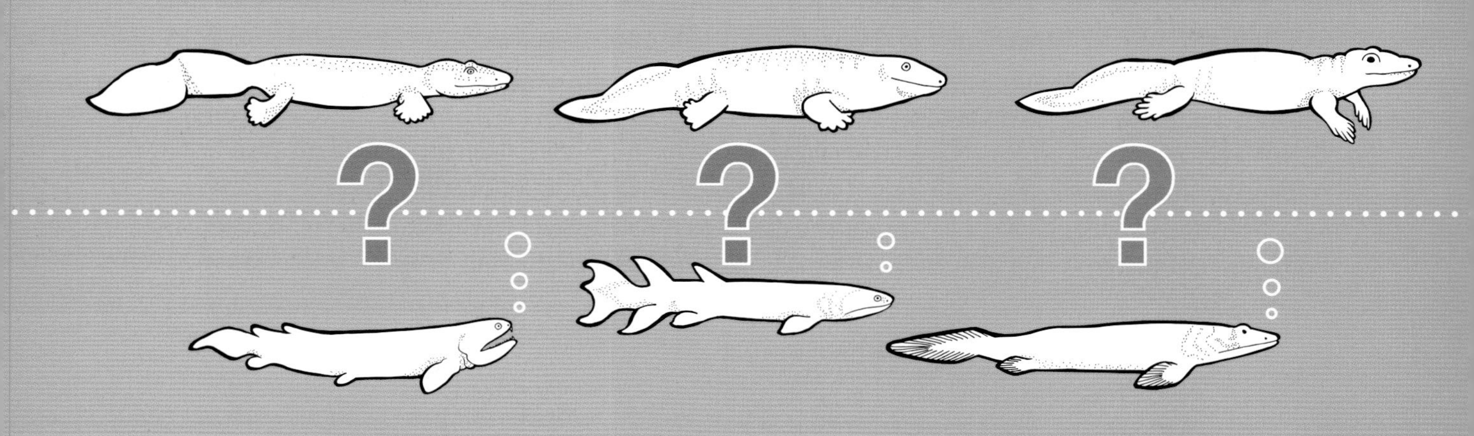

than they really are, and creates the impression of a close genealogical relationship, and an easy transition between different types of animals. Presentations of the reptile-to-mammal sequence, in particular, often enlarge some skulls and shrink others to make them appear more similar in size than they actually are.[21] *(Compare Figure 1:6 and Figure 1:8.)*

Time Problems

Textbooks also frequently fail to mention that the different skeletons shown in transitional sequences (including the mammal-like reptiles)[22] were not found close together geologically. In fact, some supposed ancestors and decendents were found in widely separated layers of sedimentary rock, representing tens of millions of years of geologic time. As zoologist Henry Gee writes, referring to fossil vertebrates in general, "The intervals of time that separate the fossils are so huge that we cannot say anything definite about their possible connection through ancestry and descent." [23]

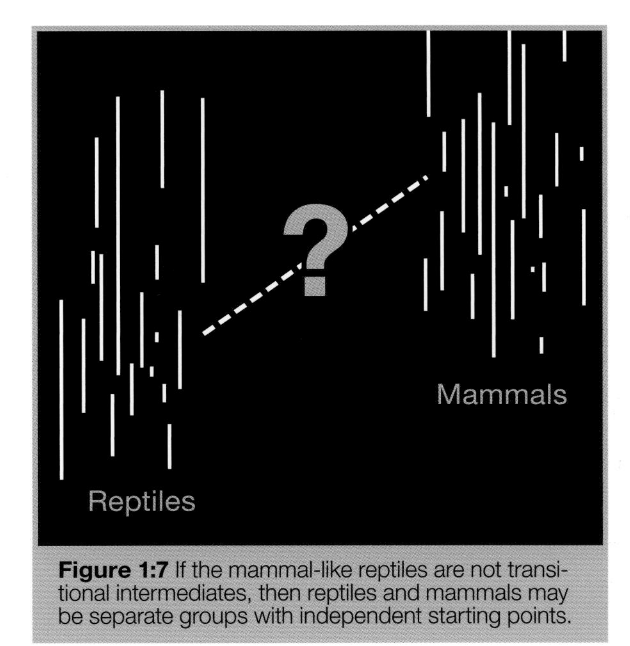

Figure 1:7 If the mammal-like reptiles are not transitional intermediates, then reptiles and mammals may be separate groups with independent starting points.

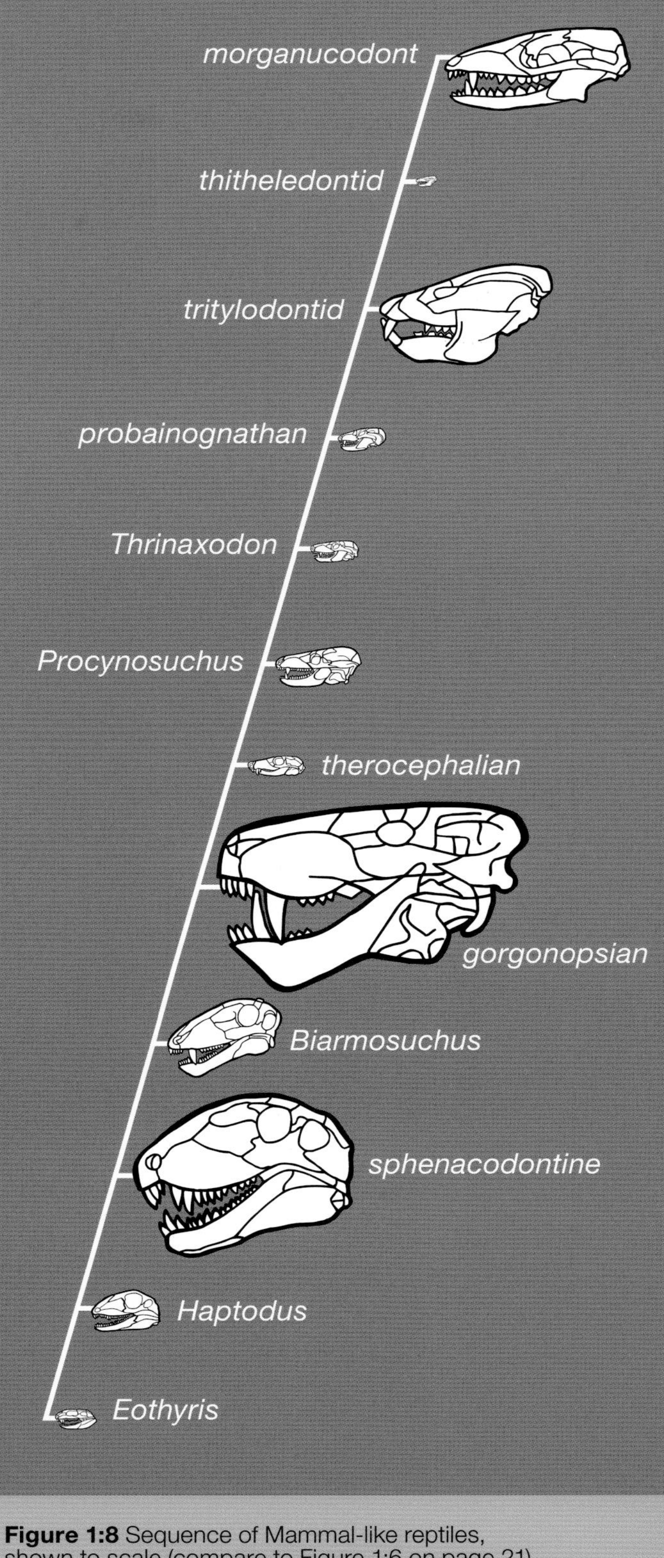

Figure 1:8 Sequence of Mammal-like reptiles, shown to scale (compare to Figure 1:6 on page 21).
Recalculated, from T.S. Kemp, The Origin & Evolution of Mammals (2005).

FOSSIL SUCCESSION: FURTHER DEBATE

Darwin himself was well aware of the problems that the fossil record posed for his theory. He spent two chapters in the ***Origin*** *discussing the fossil record, the first of which was devoted almost entirely to analyzing why the fossil evidence did not fit better with the theory of Common Descent. Where were the multitudes of transitional forms connecting different groups, as predicted (and expected) by his theory?*

According to Darwin, the answer was clear: the fossil record is "extremely imperfect." It simply failed to document transitional forms. "Nature," wrote Darwin, "may almost be said to have guarded against the frequent discovery of her transitional or linking forms."[24]

Artifact Hypothesis: Are the "Gaps" Real?

Modern neo-Darwinists have proposed an explanation for the imperfection of the fossil record, an explanation that supports Darwin's view. They call this explanation the "artifact hypothesis." According to this view, events like the Cambrian Explosion do not show that new forms of life arose suddenly. Instead, they simply show that the fossil record has been poorly sampled, and that the fossils themselves were not consistently preserved. The unfortunate and misleading result is an apparent absence of ancestors. In other words, advocates of the artifact hypothesis say that the Cambrian explosion is not real; it is only the result—or an "artifact"—of having too small a sample of fossils to work with.

Critics of the artifact hypothesis have pointed out a number of problems with this explanation. They agree that a poor (small) data sample could create an artifact, or false impression. But *is* the sample too small? Many paleontologists would argue that we have plenty of fossils. Paleontologist Mike Foote points out that, although we continue to find new fossils, those we find belong to phyla and other major groups that we already knew about.[25] This strongly suggests that the pattern of the fossil record (sudden appearance, stasis, and distinct gaps between major groups) is truly representative of the history of life, not just an artifact of poor data sampling.

As Foote concludes, "We have a representative sample... and therefore we can rely on patterns documented in the fossil record."[26]

Soft and Small?

Is it possible that the missing Precambrian transitional forms were too small to be fossilized?

Could it be that the intermediates weren't fossilized because they didn't have hard body parts like teeth or exoskeletons? Some defenders of Common Descent say yes, and point out that small structures and soft tissues are more susceptible to decay and destruction, and are, therefore, harder to preserve. This would explain why they are absent from the fossil record.

Critics agree that soft, small structures are more difficult to preserve. However, they point out that Cambrian strata around the world have yielded fossils of entirely soft-bodied animals representing several phyla.[27]

This point has been further emphasized by a recent Precambrian fossil find near Chengjiang, China. Scientists there recently discovered incredibly preserved microscopic fossils of sponge embryos. (Sponges are obviously soft-bodied. Their embryos are small and soft-bodied, too—other than their tiny spicules.) Paul Chien, a marine paleobiologist at the University of San Francisco argues that this discovery poses a grave difficulty for the artifact hypothesis. If the Precambrian rocks can preserve microscopic soft-bodied organisms, why don't they contain the ancestors to the Cambrian animals?[28] If a soft-bodied embryo can be preserved, why not an adult animal?

New "Soft" Evidence

Trace fossils provide additional evidence that the Cambrian Explosion was real. When an animal burrows through sediment, it leaves tracks or burrows behind. These tracks can be fossilized. We call these fossilized animal tracks *trace fossils*. Both soft- and hard-bodied animals could leave trace fossils. Here's the point. If lots of soft-bodied animals existed before the Cambrian, then we should find lots of trace fossils. But we don't. Precambrian sedimentary rock records very little activity. However, at the beginning of the Cambrian explosion, we see a dramatic increase in trace fossils all over the world.[29]

Punctuated Equilibrium

Many paleontologists are well aware of the conflict between the fossil record and neo-Darwinian

Statistical Sampling 101

Suppose you find a big box of marbles. You reach in and grab six marbles at random. When you remove the marbles, you discover that each marble is either red, green, or blue. Can you assume that the box contains only these colors? Not yet.

This sample is so small that it may not be representative of all the colors in the box. There could be a rainbow of intermediate colors in there. The "three colors only" hypothesis might be an artifact of a small sample.

However, you keep going until you've pulled about 1,000 marbles out of the box. You look at them all, and still find only red, green, and blue ones. There are still some marbles left in the box. What colors would you guess they are? ■

The Vendian Fossils: Was the "Cambrian Explosion" Really Explosive?

Some exciting fossil finds have revived the hope that the fossil record does, in fact, reveal evidence of Precambrian transitional forms. Four types of fossils of multicellular organisms have been found in a Precambrian stratum called the Vendian layer (565-570 million years ago) in numerous locations, including England, Newfoundland, the White Sea in northwest Russia, and the Namibian desert in southern Africa.*

The Vendian layer contains a number of strange creatures with strange names: *Kimberella, Dickinsonia,* and *Spriggina.* Some scientists have suggested that these odd creatures may well be the fossilized intermediates that neo-Darwinists have been looking for.

Other scientists contend that the Vendian fossils aren't much help in explaining the Cambrian Explosion. At best, the Vendian creatures might represent ancestral forms of a small fraction of the many new phyla that arise in the Cambrian. Except for *Kimberella,* a primitive mollusk, the body plans of these fossilized organisms have no clear relationship to any of the new organisms that appear in the Cambrian Explosion, or thereafter. Others, like *Dickinsonia* and *Springinna,* do not have eyes, mouths, or anuses, which has led many biologists to question whether these organisms were even animals. ■

*The new fossil discoveries include the strange Ediacaran fossils named for the site of the first Vendian find in the Ediacaran Hills in the Australian outback.

A. H. Knoll and S. B. Caroll, "Early animal evolution: Emerging views from camparative biology and geology," *Science* 284 (June 25, 1999):2129-2136.

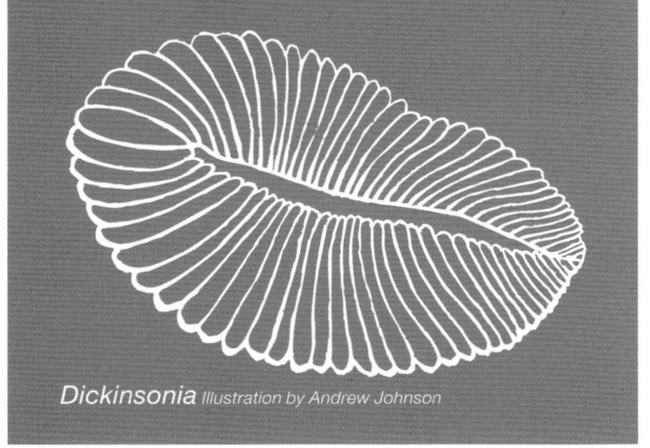

Dickinsonia Illustration by Andrew Johnson

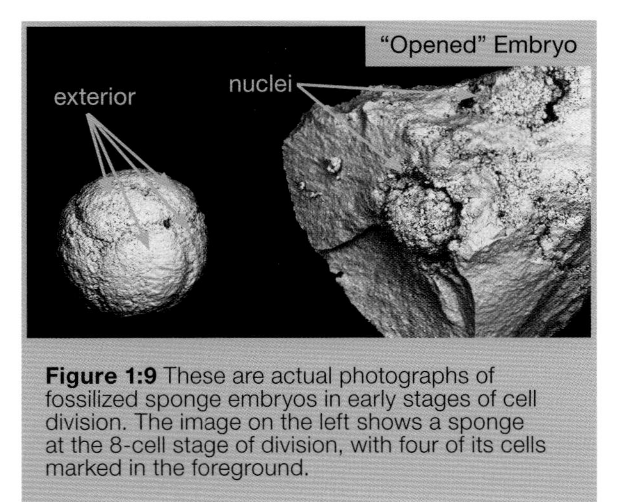

Figure 1:9 These are actual photographs of fossilized sponge embryos in early stages of cell division. The image on the left shows a sponge at the 8-cell stage of division, with four of its cells marked in the foreground.

theory. In the traditional view, the fossil record was always to blame for the missing pieces of the evolutionary puzzle. Darwin, himself, had said the fossil record was "woefully incomplete."[30]

By the early 1970s, some scientists, including paleontologists Niles Eldredge and the late Stephen Jay Gould, began to become dissatisfied with this explanation. "We paleontologists," wrote Eldredge, "have said that the history of life supports that interpretation [of gradual adaptive change], all the while really knowing that it does not."[31]

Eldredge and Gould decided to take a different approach. Instead of blaming the fossil record, they accepted the fossil data at face value. They agreed that the fossil record really does show many groups of organisms appearing abruptly, continuing unchanged for millions of years, then going extinct. "Stasis is data," they insisted.[32]

Eldredge and Gould advocated a new evolutionary theory called punctuated equilibrium,[33] a theory that says the history of life is still best represented as a single branching tree (Common Descent), but the branches split off very quickly.

This brings up an interesting question. How could the branches split off so quickly? Different advocates of punctuated equilibrium have suggested various theories of how biological change could occur rapidly.[34] One of the most popular is called *species selection.*

Species Selection

Traditional neo-Darwinism emphasizes competition among *individuals* within the *same* species. Species Selection stresses competition among rival *species*. This mechanism, say the theory's advocates, would produce rapid changes in body structure. Species selection acts on larger differences in biological form—differences between whole species, rather than differences between siblings. Evolutionary change would thus take place in bigger jumps. As a result, the branches in the Tree of Life would split off so abruptly that they would appear virtually horizontal.

Sudden species change means that fewer intermediate forms between species would have existed, and would only rarely be preserved. So, the fossil record would not preserve the "punctuation," it would only preserve the "equilibrium"—the new forms of life which remained stable over long periods of time.

Big Change in a Short Time?

Punctuated Equilibrium was immediately attractive to many paleontologists because it described the fossil record more accurately than neo-Darwinism had done.[35] However, many critics of the theory pointed out that punctuated equilibrium has never explained how the major changes recorded in the fossil record could have taken place in such a short time.[36]

Though scientists have proposed mechanisms to complement the theory of punctuated equilibrium, they have not solved this problem, critics contend. They note that species selection can at best only explain the origin of new species. It does not explain the origin of higher taxonomic groups (like phyla and classes). To describe how one species of trilobite evolved into another is not the same as explaining how trilobites arose in the first place. Nor does species selection explain the origin of new body plans or new structures (like wings and eyes) that arise with the higher taxonomic groups. Why not? Higher taxonomic groups are distinguished from one another by their distinctive traits and physical features. But what produces these new traits? Even advocates of punctuated equilibrium acknowledge that the driving force behind this process is still good old natural selection acting upon small genetic variations. But generating a phyla-level degree of biological change in this way takes time—and lots of it.

And that's the dilemma, say the critics. If the theory of Punctuated Equilibrium is right about the rate of evolutionary change—if it accurately describes how rapidly the branches of the Tree of Life split off—then it has no mechanism that can produce new structures as rapidly as the fossil record shows them arising. As invertebrate zoologist Jeffrey Levinton argues, "[I]t is inconceivable how selection among species can produce the evolution

- Some critics say neo-Darwinism is not consistent with fossil data.
- Other critics say that punctuated equilibrium is consistent with fossil evidence, but lacks an adequate mechanism.
- Critics of both views argue that there are still far fewer transitional forms in the fossil record than we would expect, even if new forms of life did arise quickly.

of detailed morphological structures...Species selection did not form an eye."[37]

Eventually, even advocates of punctuated equilibrium were forced to agree. "I recognize," wrote Gould, "that we know no mechanism for the origin of such organismal features other than conventional natural selection at the organismic level...."[38]

One Final Dilemma

This has lead to a kind of Catch-22. Neo-Darwinists have critiqued punctuated equilibrium because they say it lacks a mechanism that can produce biological change as fast as the fossil record requires. On the other hand, advocates of punctuated equilibrium have critiqued neo-Darwinism because the fossil record contradicts the neo-Darwinian picture of the history of life. Critics of both argue that there are far fewer transitional forms in the fossil record than we would expect even if new forms of life arose quickly.

Where Does This Leave Us?

We have seen that scientists disagree over how to interpret the fossil evidence.[39] Some see evidence of an orchard of separate trees; others see a single, continuous, branching tree. But how can there be disagreements? Facts are facts, right? How can qualified scientists disagree over evidence? ***Let's look at a real-life example.***

Figure 1:10 is a reproduction of a fossil exhibit that was on display at the California Academy of Sciences in the 1990s. It showed fossils arranged in the familiar branching-tree pattern.[40] The dots on the display represented fossils from several distinct phyla, or "body plans." Notice that the phyla lines are parallel, illustrating that each phylum remains distinct—separate from other phyla—during the entire time it appears in the fossil record. This exhibit contained many different specimens of each separate phylum.

You should also notice that the phyla lines in this exhibit converged, or came together, toward the bottom of the display. Why? Because according to the Theory of Universal Common Descent, there were transitional intermediate forms that bridged the gaps between these distinct animal groups early in the history of life.

The display didn't show any pictures of these forms, however. Instead, it represented

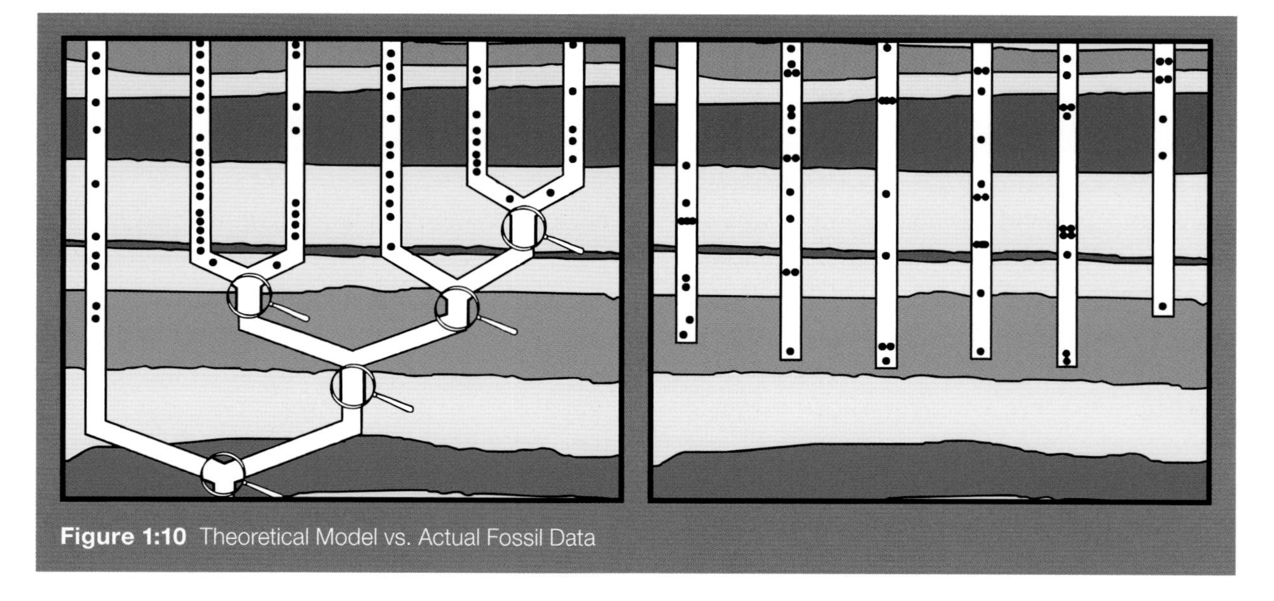

Figure 1:10 Theoretical Model vs. Actual Fossil Data

each connecting point with a large magnifying glass. Interestingly, if you had looked carefully through the magnifying glass, you wouldn't have seen anything underneath. Why? These connecting forms have not been found in the fossil record.[41]

These are the facts. Now, where's the disagreement?

Scientists, like detectives, do more than simply collect facts. They also interpret facts. They try to put facts together to create a more complete story about their subject. This disagreement is over how the facts of the fossil record should affect the story we tell about life's history.

For example *(see Figure 1:11)*, some scientists say the absence of transitional forms should dramatically change the story we tell about life's history. They point out that when we study the fossils we have actually found, the evidence does not lead us to connect the major lines of descent into a single branching tree.

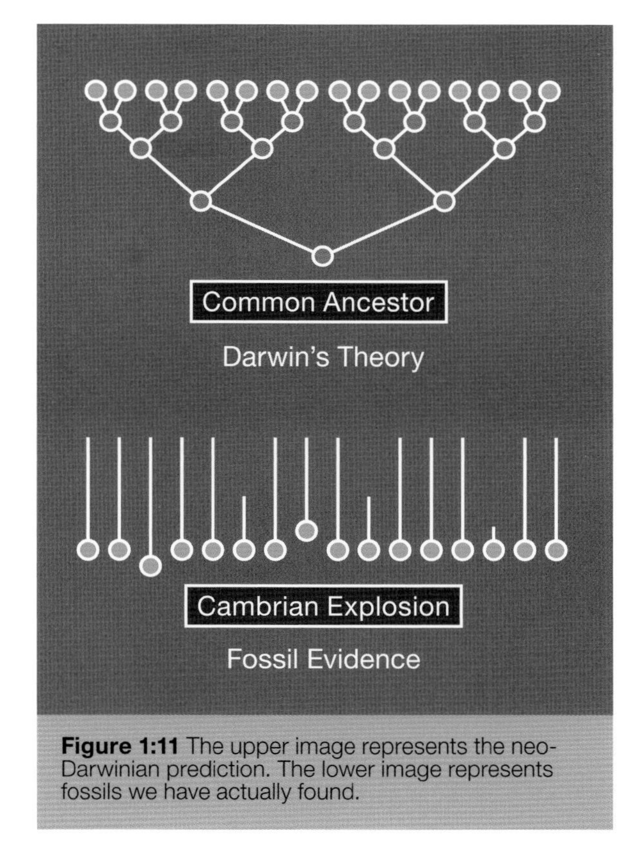

Figure 1:11 The upper image represents the neo-Darwinian prediction. The lower image represents fossils we have actually found.

In contrast, advocates of Common Descent say the absence of fossil intermediates is not necessarily a problem for the Darwinian story. Why? The fossil record does show evidence of some transitional forms (*Archaeopteryx* and the mammal-like reptiles). This, they say, makes it reasonable to think that others will be found.

The absence of fossil intermediates is not necessarily a major concern for advocates of Punctuated Equilibrium either, at least at the species level. According to this version of the story, intermediate forms are naturally hard to find because evolutionary changes between species occur rapidly. Rapid change means fewer transitional forms to start with. In other words, fossils of transitionals are out there, but the chance of unearthing them is simply a needle-in-a-haystack proposition. Even so, some advocates of punctuated equilibrium *do* acknowledge that the absence of transitions between major groups of organisms is an unsolved problem for evolutionary theory as a whole.

The Debate Goes On

Is Darwin right when he says that nature has "guarded against the discovery of the transitional forms?" If so, did evolution occur slowly and gradually, as Darwin thought, or in rapid bursts, as advocates of punctuated equilibrium think? Or, are the critics of Universal Common Descent right when they say that transitional forms have not been discovered because they never existed? Will genuine transitional forms be found? Have they already been found? Which of the three pictures of the history of life is correct *(see Figure 1:12)*? Scientists disagree, and the issue is far from settled.

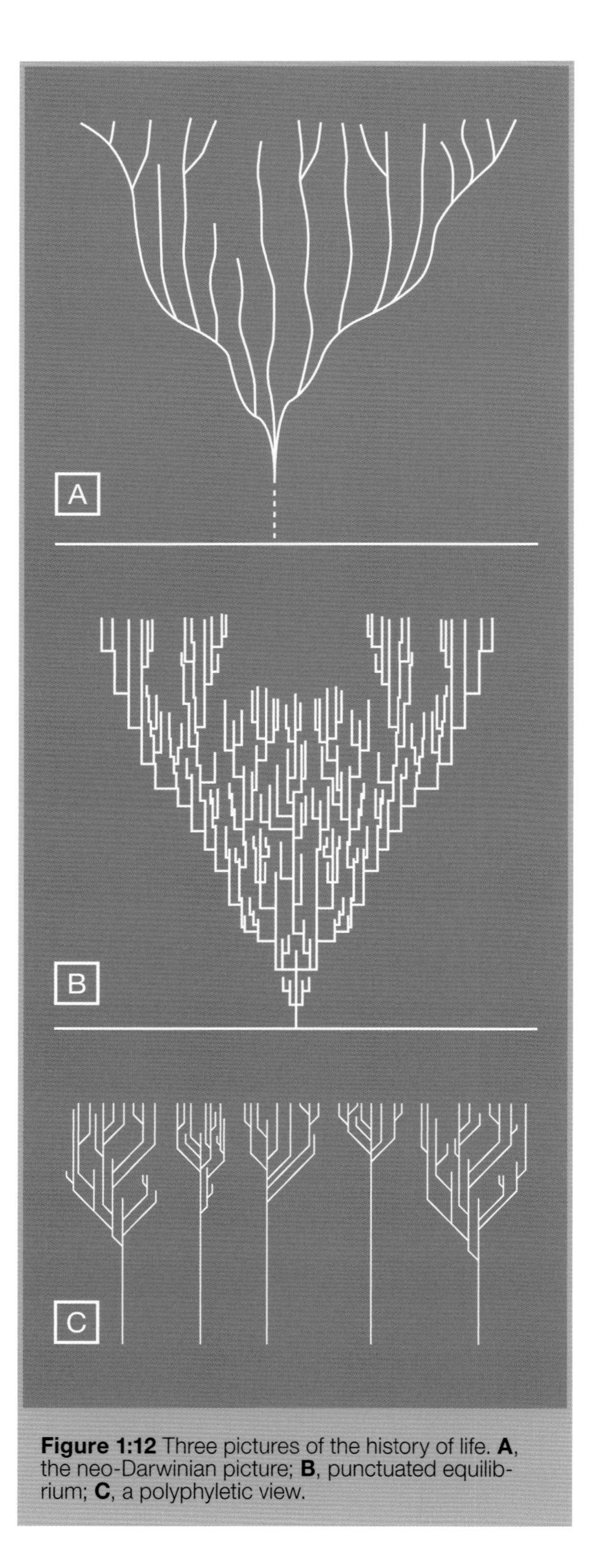

Figure 1:12 Three pictures of the history of life. **A**, the neo-Darwinian picture; **B**, punctuated equilibrium; **C**, a polyphyletic view.

By the way, there is another important reason that advocates of Universal Common Descent aren't bothered by the absence of fossil intermediates. They think there is a lot of other evidence that supports their theory. In the next four chapters, we will examine some of this evidence, and we will see what both supporters and critics of Common Descent have to say about it.

Endnotes

1 In the Recapitulation and Conclusion chapter of the 2nd edition of the Origin, Darwin stated "...probably all the organic beings which have ever lived on this earth have descended from some one primordial form, into which life was first breathed." Charles Darwin, *On the Origin of Species* (Cambridge, Mass: Harvard University Press, 1964 [Facsimile of the First Edition, 1859]):484.

2 Philip Gingerich, "The Whales of Tethys," *Natural History* 103 (April 1994):86-88.

C. Zimmer, "Back to the sea," *Discover* 16 (1995):82-84.

P. Gingerich, S.M. Raza, M. Arif, M. Anwar, X. Zhou, "New Whale From the Eocene of Pakistan and the Origin of Cetacean Swimming," *Nature* 368 (1994):844-847.

P. Gingerich, B.H. Smith, E.L. Simons, "Hind Limbs of Eocene Basilosaurus: evidence of feet in whales," *Science* 249 (1990):154-157.

3 Adrian Desmond and James Moore, *The Life of a Tormented Evolutionist: Darwin*. (New York, NY: Warner Books, Inc. 1992):514.

4 Robert L. Carroll, "Towards a new evolutionary synthesis," *Trends in Ecology and Evolution* 15 (January 2000):27-32.

Everett C. Olson, "The problem of missing links: today and yesterday," *The Quarterly Review of Biology* 56 (1981):405-442.

5 James W. Valentine, *On the Origin of Phyla* (Chicago: University of Chicago Press, 2004):35. "[M]any of the branches [of the Tree of Life], large as well as small, are cryptogenetic (cannot be traced into ancestors). Some of these gaps are surely caused by the incompleteness of the fossil record (chap.5), but that cannot be the sole explanation for the cryptogenetic nature of some families, many invertebrate orders, all invertebrate classes, and all metazoan phyla."

6 This is also surprising because we find the Cambrian explosion past midfield, all the way down to about the 12-yard line. This is much farther 'downfield' than one might have supposed. This sudden, brief explosion contradicts Darwin's image of a gradually branching tree.

see also James W. Valentine and David Jablonski, "Morphological and developmental macroevolution: a paleontological perspective," *International Journal of Developmental Biology* 47 (2003):517-522.

7 S.A. Bowring, J.P. Grotzinger, C.E. Isachsen, A.H. Knoll, S.M. Pelechaty, & P. Lolosov, "Calibrating rates of early Cambrian evolution," *Science* 261 (1993):1293-1298.

S.A. Bowring, J.P. Grotzinger, C.E. Isachsen, A.H. Knoll, S.M. Pelechaty, & P. Lolosov, "A new look at evolutionary rates in deep time: Uniting paleontology and high-precision geochronology," *GSA Today* 8 (1998):1-8.

S. Aris-Brosou and Z. Yang, "Bayesian models of episodic evolution support a late Precambrian explosive diversification of the Metazoa," *Molecular Biology and Evolution* 20 (2003):1947-1954.

Robert L. Carroll, "Towards a new evolutionary synthesis," *Trends in Ecology and Evolution* 15 (January 2000):27-32.

8 Susumo Ohno, "The notion of the Cambrian pananamalia genome," *Proceedings of the National Academy of Sciences* 93 (Aug 1996):8475-8478.

9 Nancy B. Simmons, "An Eocene Big Bang for bats," *Science* 307 (January 28, 2005):527-528.

10 Stefanie De Bodt, Steven Maere, and Yves Van de Peer, "Genome duplication and the origin of angiosperms," *Trends in Ecology* and *Evolution* 20 (2005):591-597. "In spite of much research and analyses of different sources of data (e.g., fossil record and phylogenetic analyses using molecular and morphological characters), the origin of the angiosperms remains unclear. Angiosperms appear rather suddenly in the fossil record... with no obvious ancestors for a period of 80-90 million years before their appearance."

see also Charles B. Beck, *Origin and Early Evolution of Angiosperms* (New York: Columbia University Press, 1976):5.

11 Correspondence between Charles Darwin and his friend Joseph Hooker. "The rapid development as far as we can judge of all the higher plants within recent geological times is an abominable mystery." Quoted by Henry Dunkinfield Scott, *The Evolution of Plants* (New York: Henry Holt, 1911):37.

12 The full quotation reads, "The case at present must remain inexplicable; and may be truly urged as a valid argument against the views here entertained." Charles Darwin, *On the Origin of Species* (Cambridge, Mass: Harvard University Press, 1964 [Facsimile of the First Edition, 1859]):308.

13 Technically, to say that phyla remain stable is almost redundant. After all, scientists define phyla by referring to an unchanging set of anatomical characteristics. In another sense, however, the stability of phyla is remarkable.

Think of the different phyla as though they were arranged like bars on a bar chart. Each bar represents a unique body plan. The farther apart two individual bars are from one another, the more different the anatomical characteristics are.

In nature, an animal body plan could theoretically fall anywhere along this continuum, even in the gaps between bars. But that's not what happens. Individual animals either falls within one of the existing phyla, or in some instances new animals are found that represent radically new body plans altogether. Phyla, as descriptions, remain stable by definition. What's remarkable is that reality—the forms of actual animals representing the phyla—remains stable, too.

14 David M. Raup, "Conflicts between Darwin and paleontology," *Field Museum of Natural History Bulletin* 50 (1979):22-29.

15 Joel Cracraft, "Phylogenetic Analysis, Evolutionary Models, and Paleontology," in *Phylogenetic Analysis and Paleontology*, eds. Joel Cracraft and Niles Eldredge (New York: Columbia University Press, 1979):7-39. Cracraft goes on to say, "Indeed, the 'factual information' that Darwin presents (there was virtually none) seems to support a philosophical (and scientific) viewpoint opposite to that of his own. Darwin was the consummate theorist, a scientist of the highest stature who did not let data stand in the way of his ideas."

16 Giuseppe Sermonti and Roberto Fondi, *Dopo Darwin: Critica al' Evoluzionismo* (Milan: Rusconi, 1980).

Stephen C. Meyer, Marcus Ross, Paul Nelson, and Paul Chien, "The Cambrian Explosion: Biology's Big Bang," in *Darwinism, Design, and Public Education*, eds. John Angus Campbell and Stephen C. Meyer (East Lansing: Michigan State University Press, 2003):323-402.

Malcolm S. Gordon and Everett C. Olson, *Invasions of the Land: The Transitions of Organisms from Aquatic to Terrestrial Life* (New York: Columbia University Press, 1995):128-133, 262-264.

17 R.L. Carroll, "Limits to knowledge of the fossil record," *Zoology* 100 (1997/98):221-231.

18 T.S. Kemp, *Mammal-Like Reptiles and the Origin of Mammals,* (London, England: Academic Press, 1982):296.

More recently, Kemp restates his view. "As it happens, the fossil record of the mammal-like reptiles is still by far the best paleontological documentation of the origin of a major new taxon, notwithstanding recent discoveries of transitional grades of tetrapods...and birds." T.S. Kemp, *The Origin & Evolution of Mammals* (New York: Oxford University Press, 2005).

19 Mark A. Norell, Michael J. Novacek, "The fossil record and evolution: comparing cladistic and paleontologic evidence for vertebrate history," *Science* 255 (Mar 27, 1992):1690-1693.

20 Mark A. Norell, Michael J. Novacek, op.cit.

21 Some authors do include the scaling ratios they use, leaving it up to the audience's mathematical skills to calculate actual comparative size. Other authors use a scale legend line, and it's up to the reader to notice that the same length line that represented 2 cm in Picture A represents 10 cm in Picture B. Still other authors simply put "Skulls not to scale," somewhere in the caption. Unless students read the fine print and do the calculations, they are often left with a very misleading impression of the similarity of the animals in these alleged sequences.

22 T.S. Kemp, *The Origin & Evolution of Mammals* (New York: Oxford University Press, 2005).

23 Henry Gee, *In Search of Deep Time: Beyond the Fossil Record to a New History of Life* (New York: The Free Press, 1999):23.

24 Charles Darwin, *On the Origin of Species* (Cambridge, Mass: Harvard University Press, 1964 [Facsimile of the First Edition, 1859]):292.

25 Mike Foote, "Sampling, taxonomic description, and our evolving knowledge of morphological diversity," *Paleobiology* 23 (Spring 1997): 181-206.

26 Mike Foote, op. cit.

27 James W. Valentine, "The Macroevolution of Phyla," *in Origin and Early Evolution of the Metazoa,* ed. Jere H. Lipps and Phillip W. Signor (New York: Plenum Press, 1992) 525-53. Also, see section 3.2, "Soft-bodied Fossils":529-31.

28 Paul Chien, J.Y. Chen, C.W. Li, Frederick Leung, "SEM Observation of Precambrian Sponge Embryos From Southern China, Revealing Ultrastructures Including Yolk Granules, Secretion Granules, Cytoskeleton and Nuclei." Paper presented to North American Paleontological Convention, University of California, Berkeley, June 26-July 1, 2001.

see also James W. Hagadorn et. al., "Cellular and subcellular structure of neoproterozoic animal embryos," *Science* 314 (October 2006):291-294.

29 James W. Valentine, *On the Origin of Phyla* (Chicago: University of Chicago Press, 2004):179-80.

James W. Valentine, David Jablonski, and Douglas H. Erwin, "Fossils, molecules and embryos: new perspectives on the Cambrian Explosion," *Development* 126 (1999):851-859; pp. 853-53.

Mary L. Droser, Sören Jensen, and James G. Gehling, "Trace fossils and substrates of the terminal Proterozoic-Cambrian transition: Implications for the record of early bilaterians and sediment mixing," *Proceedings of the National Academy of Sciences* 99 (2002):12572-12576; p. 12575.

30 Charles Darwin, *On the Origin of Species By Means of Natural Selection or The Preservation of Favoured Races in the Struggle for Life* (Cambridge, Mass: Harvard University Press, 1964 [Facsimile of the First Edition, 1859]).

31 Niles Eldredge, *Time Frames: The Rethinking of Darwinian Evolution and the Theory of Punctuated Equilibria* (New York: Simon & Schuster, 1985):144.

32 Stephen Jay Gould and Niles Eldredge, "Punctuated equilibrium comes of age," *Nature* 366 (November 1993):223.

33 Niles Eldredge and Stephen Jay Gould, "Punctuated Equilibria: An Alternative To Phyletic Gradualism" in *T.J.M. Schopf Models In Paleobiology* (San Francisco: Freeman, Cooper and Company, 1972): 82 -115.

34 Advocates of punctuated equilibrium suggested that a mechanism known as allopatric speciation could explain how biological change could arise very rapidly. The prefix "allo-" means other or different, and the suffix "-patric" means father. Thus, allopatric speciation means generating new species from separate populations (different fathers).

Here's how it works. Imagine a population of mammals all living in the same area. (We'll call them Population A.) One day, a flash flood sends a torrent of water right down the middle of their habitat. The flash flood changes the course of a nearby river, and Population A is now split into two groups, physically separated by the river. (It has become Population B and Population C.) Over time, Population B migrates down into an area of grassier meadows, while Population C migrates up into the rockier hill country. Over time, each population adapts to its new environment. A long time later, another change in the environment forces Population B and Population C into their original habitat.

What will paleontologists find when they study the strata of the original habitat? They will see Population A remaining stable for a long time. Then suddenly (geologically speaking), Population A will have disappeared, and Population B and Population C will have appeared. The fossil strata appear to record a rapid change, even though the change really took place gradually, somewhere else. Advocates of allopatric speciation argue that this is a sensible explanation for both the fossil data and for the apparent rapid emergence of new species.

35 D.V. Ager, "The nature of the fossil record," *Proceedings of the Geological Association* 87 (1976):131-159.

36 In fairness, advocates of punctuated equilibrium never intended to describe or explain evolutionary changes above the species level. Yet, for just this reason, the theory does not solve the problem of the appearance origin of major new forms of life, for example, those we see in the Cambrian explosion.

37 Jeffrey Levinton, *Genetics, Paleontology, and Macroevolution* (Cambridge: Cambridge University Press, 1998):208.

38 Stephen Jay Gould, *The Structure of Evolutionary Theory,* (Cambridge: Harvard University Press, 2002):710.

39 Everett C. Olson, "The problem of missing links: today and yesterday," *Quarterly Review of Biology* 56 (December 1981):405-441.

40 The picture depicts a museum display called the "Hard Facts Wall," part of the Exhibit "Life Through Time: The Evidence for Evolution," put on display by the California Academy of Sciences in the 1990s.

41 If you had looked closely at the Hard Facts Wall, you might have noticed an interesting discrepancy. The exhibit actually showed a 440-million-year-old fossil as being older than a 570-million-year-old fossil. Why? Maybe to make the display look more like a single tree? Critics say the reason the exhibit looked like a single tree was that the people who set it up were convinced ahead of time that it would look like a single tree.

Anatomical Homology

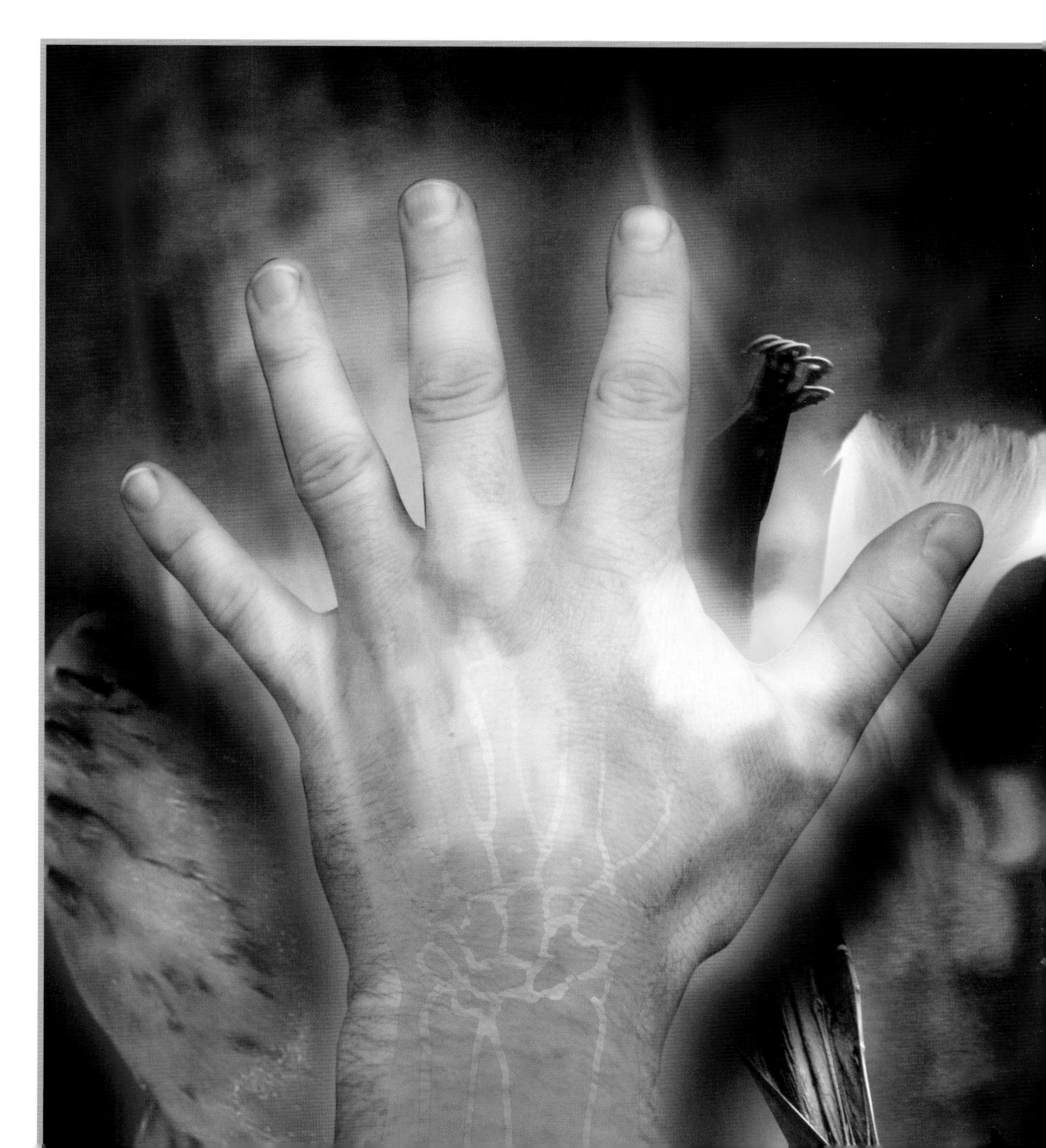

ANATOMICAL HOMOLOGY: CASE FOR

You've seen that there is a spirited debate over what the fossil record actually tells us about the history of life. Scientists are also debating another class of evidence. When they compare the anatomy of different animals, they find that many of their body parts are remarkably similar. This class of evidence is known as Homology.

If you've ever dissected a pig, you know that its forelimb has three large bones, the humerus, radius, and ulna. Its front hoof is made up of a host of smaller bones. If you look at an x-ray image of your own arm, you'll see a similar pattern: humerus, radius, and ulna, followed by the many smaller bones of your wrist and hand *(see Figure 2:1)*. This raises an interesting question. Why should the pig's forelimb and your own arm have the same skeletal pattern? After all, the pig uses its forelimbs mainly for locomotion—getting around the pen. You use your arms and hands for a wide range of tasks, such as catching a baseball, typing a report, or lifting a box—but not for locomotion. Many biologists before Darwin thought that these similarities (called "homologies") were due to a common plan or "archetype." But Darwin rejected this idea. Instead, he said these homologies were best explained by his theory of descent from a common ancestor.[1]

According to Darwin (and modern evolutionary biologists), homologies exist because organisms inherited these structures from the ancestor they had in common. The human forelimb resembles that of the pig because humans and pigs are related, and are descended from a common ancestor that also possessed the humerus-radius-ulna skeletal pattern.

Of course, pigs and humans did not inherit the bones themselves from their common ancestor. That would be ludicrous.* So, what is it that "descends" from ancestor to descendent? What actually passes from parent to offspring is the genetic information carried in the parents' egg and sperm. This information guides the construction of the forelimb (as well as the rest of the body) as pigs and humans develop from embryo to adulthood.

Let's say we have two different animals, A and B, which have homologous structures. According to the theory of Common Descent (Evolution #2), A and B have these homologous structures because they were built by homolo-

* Not to mention borderline disgusting.

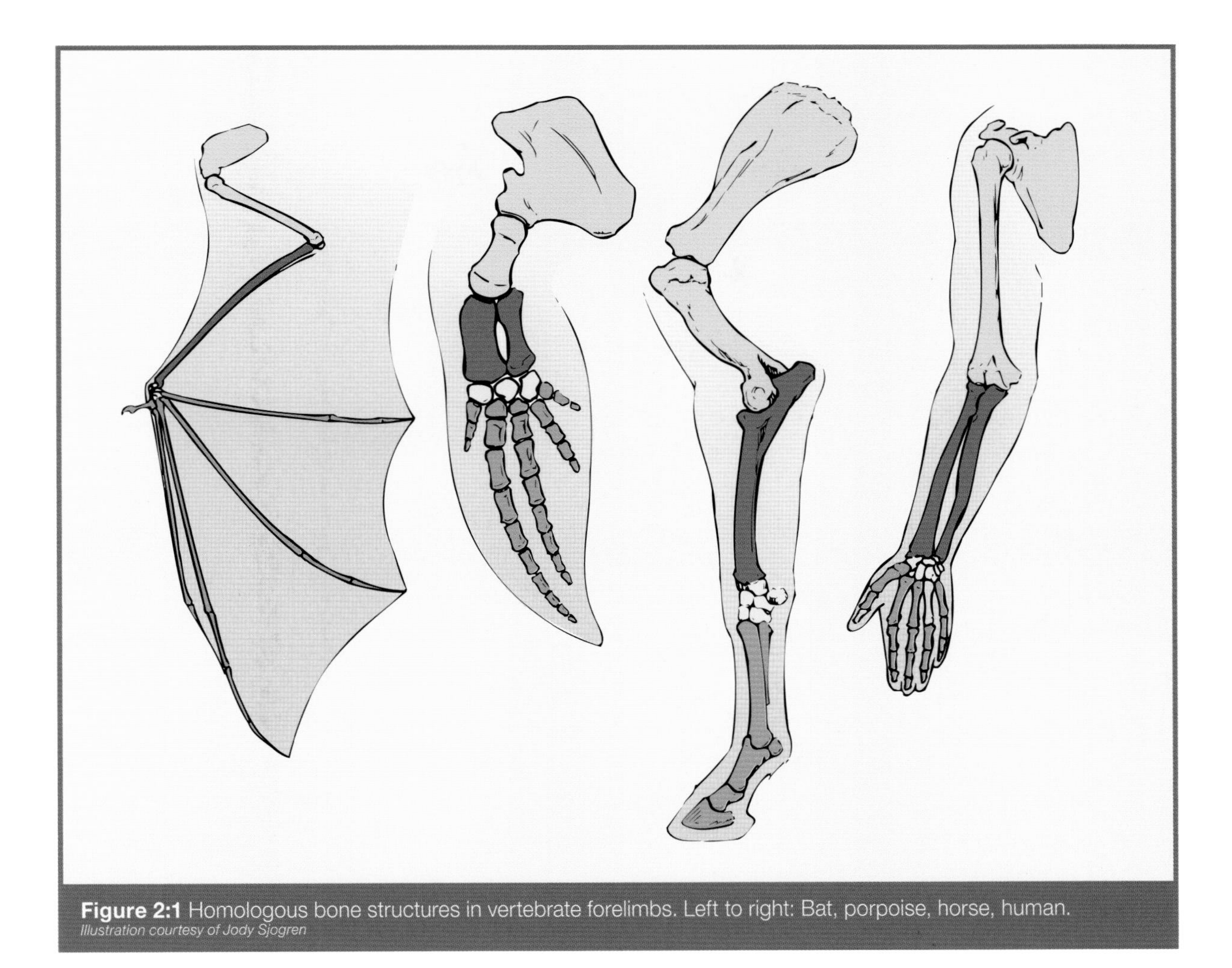

Figure 2:1 Homologous bone structures in vertebrate forelimbs. Left to right: Bat, porpoise, horse, human.
Illustration courtesy of Jody Sjogren

gous genes. Further, the processes by which the embryos of A and B become adults (called their "developmental pathways") are homologous. Why? Because both A and B descended from a common ancestor.

"In reality every organ is re-created anew in every generation," say the neo-Darwinian biologists Alfred Romer and Thomas Parsons. "[T]he identity between homologues is based upon the identity or similarity of the developmental properties... [and] hereditary units, the genes."[2]

According to neo-Darwinism, natural selection modifies genes over time. It also modifies the embryological processes that produce anatomical structures (like the forelimbs of vertebrates). As a result of these modifications, related organisms have anatomical structures that are mostly similar, with a few differences here and there. Since this is precisely what biologists see, advocates of Common Descent conclude that homologous anatomical structures originated in a common ancestor, and were then modified by natural selection.

They also say that natural selection has modified the shapes and sizes of bones—in a few cases, eliminating some entirely. Even so, we can still see the original ancestral pattern in those structures we recognize today as being similar.

"Common ancestry is not in dispute, not because of fossils, but because of features we share thanks to our common evolutionary birthright."
Henry Gee, In Search of Deep Time *(New York: The Free Press, 1999):35.*

Of course, skeletal patterns are not the only homologies. Evolutionary biologists see homology in many other anatomical structures, including soft-tissue structures like the brain and nervous system, blood vessels, muscles, and the digestive system. They say that these similarities provide further evidence of a long process of evolutionary change, or what Darwin called "descent with modification from a common ancestor."

Homologies: What and How

- A genetic program encoded in DNA directs the development of the embryo.
- The process of reproduction passes this program from one generation to the next.**
- Sometimes a section of DNA is copied incorrectly, which modifies the program.
- As a result, descendants of the original organism may have structures that are similar, but not identical, to the original.

Anatomical Homology: Reply

Other scientists have a different view of the evidence from homology. Critics of this argument say that just because some organisms have similar structures doesn't necessarily mean that all organisms had a common ancestor. Homologous structures can also be explained in other ways. Critics also argue that some evidence contradicts the Darwinian interpretation of homology.

Biologists before Darwin's time, including the French anatomist Georges Cuvier and the Harvard paleontologist Louis Agassiz, knew about the existence of homologous structures. However, they rejected common ancestry as the explanation for these similarities.

Common Features for Common Function

Agassiz, for example, explained homologies as the result of the necessity of using similar structures to solve similar functional problems.[3] On this view, the pattern we see in the vertebrate forelimb—a single bone closest to the trunk, two bones in the next segment, and a variety of bones in the segment farthest out—exists for important functional reasons. *(See pages 46-47)* Think of an example from human technology. Cars and airplanes both have wheels, not because airplanes evolved from cars, but because they both need to move easily over horizontal surfaces.

Homology Due to Natural Laws

Some modern biologists explain homology in another way. Brian Goodwin of the Open University says homology does not reflect a process of historical change, but instead reflects constraints imposed upon the structure of organisms by the laws of nature. Goodwin contends that the laws of nature dictate that a liquid, for example, has only a limited number of shapes it can take—a spiraling funnel when going down the drain, a droplet when it falls, and so on. In the same way, says Goodwin, the laws of nature ensure that only a certain number of anatomical patterns are possible.[4] Therefore, we should expect to see similarities in the anatomical structure of even different types of organisms.

** Remember what descends from parent to offspring. Romer and Parsons remind us that a boy does not inherit his skull from his parents like some kind of heirloom. Rather, he inherits the genetic information stored in the parents' sex cells (egg and sperm). This information guides the construction of his skull—and the rest of his body—as he develops from embryo to adulthood. Alfred S. Romer, Thomas S. Parsons, *The Vertebrate Body* (Philadelphia, PA: W.B. Saunders, 1977):9.

Taking A Closer Look

We have seen that some scientists have tried to explain homology apart from Common Descent. Some say similarity is due to common functional requirements. Others say similarity is due to natural laws. These scientists argue that their theories can explain the evidence as well as Common Descent.

Now, we will see that other scientists simply dispute the neo-Darwinian explanation of homology. They contend that there are important facts about homologous structures that Common Descent cannot explain.

They point out that when two or more adult structures appear to be homologous, neo-Darwinism tells us that those structures should have been built by homologous developmental pathways and homologous genes.[5] (Remember that the "developmental pathway" is the process by which the embryo physically becomes an adult.)

Contrary to these predictions, biologists are learning that homologous structures can be produced by different genes and may follow different developmental pathways.[6]

Consider fruit flies and wasps, for example. Most biologists think the adult body segments of these insects are homologous. In the neo-Darwinian view, such homologous structures should have arisen from homologous genes and homologous developmental pathways. Why? Because the genes and pathways that produce homologous structures should have been inherited from a common ancestor. Yet, surprisingly, the body segments of some wasps arise from developmental pathways that are entirely different from those of fruit flies, and even from other wasps.[7]

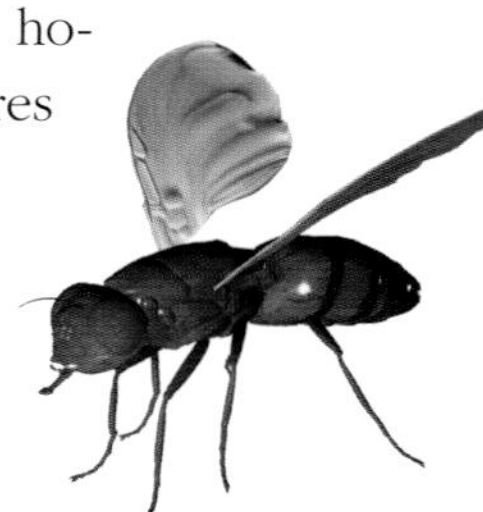

The vertebrate gut is another example of a homologous structure that develops in many different ways. In sharks, for example, the gut develops from cells in the roof of the embryonic cavity. In lampreys, the gut develops from cells on the floor of the cavity. And in frogs, the gut develops from cells from both the roof *and* the floor of the embryonic cavity. This discovery—that homologous structures can be produced by different developmental pathways—contradicts what we would expect to find if all vertebrates share a common ancestor.

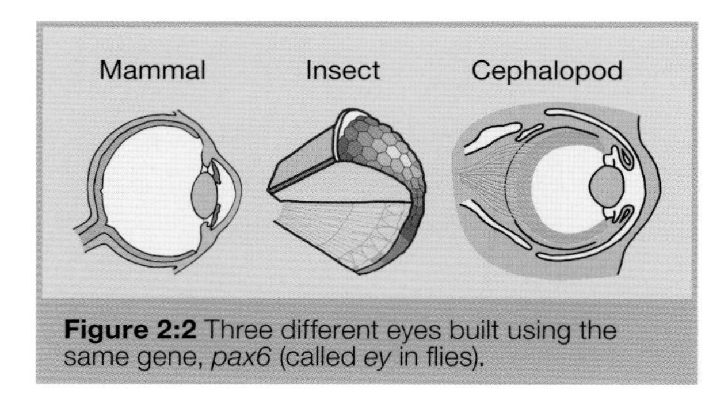

Figure 2:2 Three different eyes built using the same gene, *pax6* (called *ey* in flies).

In another surprising twist, biologists have also discovered many cases in which the *same* genes help to produce *different* adult structures.[8] Consider, for instance, the eyes of the squid, the fruit fly, and mouse. *(see Figure 2:2)* The fruit fly has a compound eye, with dozens of separate lenses. The squid and mouse both have single-lens camera eyes, but they develop along very different pathways, and are wired differently from each other. Yet the same gene is involved in the development of all three of these eyes.

According to neo-Darwinian theory, the development of non-homologous structures should be regulated by non-homologous genes. So, many biologists were taken aback to discover that non-homologous eyes (from an insect, a mollusk, and a vertebrate) could be regulated

> *"What mechanism can it be that results in the production of homologous organs, the same 'patterns,' in spite of their not being controlled by the same genes? I asked this question in 1938, and it has not been answered."*
>
> *Evolutionary biologist Gavin de Beer,* Homology: An Unsolved Problem *(Oxford: Oxford University Press, 1971).*

during their development by homologous genes, like *Pax-6*. Stephen Jay Gould called this discovery, "...unexpected under usual views of evolution...."[9]

To summarize, biologists have made two discoveries that challenge the argument from anatomical homology. The first is that the development of homologous structures can be governed by different genes and can follow different developmental pathways. The second discovery, conversely, is that sometimes the same gene plays a role in producing different adult structures. Both of these discoveries seem to contradict neo-Darwinian expectations.

Convergence?

Remember that evolutionary biologists define homology as "similarity due to common ancestry." This brings up an interesting question: are there some similarities *not* due to common ancestry? Surprisingly, nearly all biologists say there are.

Look at *Figure 2:3*. The flippers of a whale and an ichthyosaur have very similar shapes, even though the whale is a mammal and the

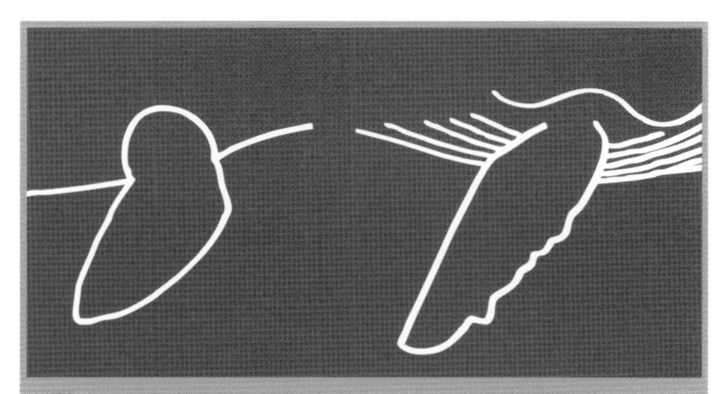

Figure 2:3 Ichthyosaur and Whale: compare the shape of the flippers. *Illustration by Janet Moneymaker*

ichthyosaur was a reptile. *Figure 2:4* shows another example. The forelimb of a mole cricket is very similar to a mole's forelimb, even though the mole is a mammal and the mole cricket is an insect.

Even biologists who take a monophyletic view of life's history will tell you that the similarity we see in these structures is not the result of common ancestry. They contend that the last common ancestor of these creatures did not possess the similar structure. In other words, the similar structures arose separately on independent lines of descent. *{cont. on page 48}*

Figure 2:4 Forelimb of the Mole Cricket (top) and a Mole (bottom) *Mole Cricket Photo by Thomas Walker; Mole Photo by Charles Schurch Lewallen*

Give me a hand, will you?

Look at the various limbs shown in *Figure 2:1.* At first glance, these limbs look very different from each other—the long, slender digits of the bat's wing, for instance, versus the short, sturdy digits of the turtle forelimb. However, they do have something very basic in common.

They all have the same basic arrangement of the major bones. Each of these creatures has:

- one bone, the humerus, in the part of the limb closest to the trunk (body);
- two bones, the radius and ulna, in the next portion of the limb; and
- a variety of smaller bones in the part of the limb farthest from the trunk—the wrist and hand (or "manus").

Let's take a closer look at the humerus-radius-ulna pattern: the one-to-two pattern shared by all the limbs. According to Charles Darwin, it doesn't make any sense to claim that these animals have similar limbs because it's a functional necessity for vertebrates with limbs to have one bone nearest their body. "Nothing can be more hopeless," he said, "than to attempt to explain this pattern of similarity in members of the same class, by [their usefulness]." In other words, this similarity of arrangement does not exist because it was planned for the tasks these organisms face. Rather, the similarities exist simply because of history. They have a common arrangement because they have a common ancestor.

Many modern evolutionary biologists agree with the "history, not utility" position. For instance, evolutionary geneticist Francisco Ayala writes, "An engineer could design better limbs in each case."* Are Darwin and Ayala right? Is it "hopeless" to look for usefulness? Let's try a thought experiment.

Look at the hypothetical limbs depicted below. Consider the "two-to-one" limb, for instance. It has two bones nearest the trunk, and one in the next limb segment—reversing the usual one-to-two pattern.

Now, right off the top of your head: can you

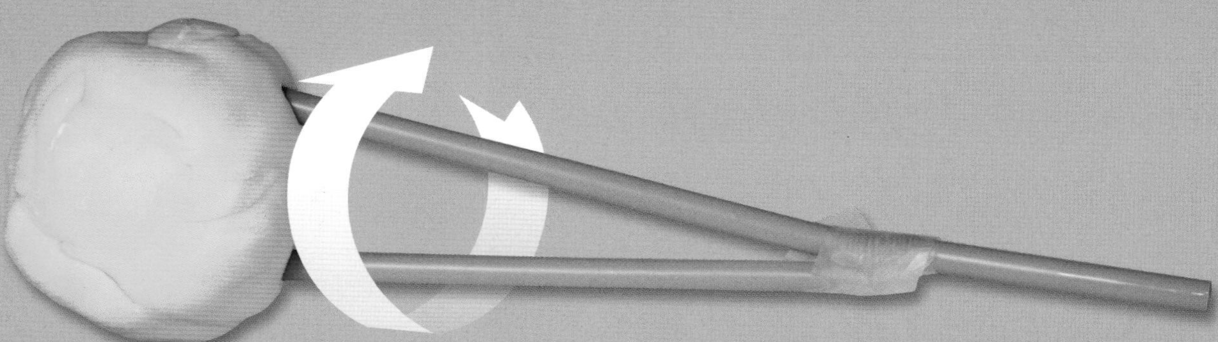

two-to-one

think of any functional difficulties that might arise from this two-to-one arrangement? Do you see any practical problems this creature might face that wouldn't arise with the familiar one-to-two pattern?

Let's think about the range of motion in the "two-to-one" limb. Would it be lesser or greater than it would be with the one-two pattern? You can answer this question for yourself. Attach a pair of equal-length straws to a lump of clay. Tape their free ends together. Now, rotate the point where you've connected them, and note how far it can move without either straw detaching from the clay.

Now connect a single straw of the same length to the clay. Notice how far its free end can move without the embedded end detaching from the clay.

Do you see the functional problems with the two-to-one pattern? Clearly, it limits the range of motion. There are obvious functional advantages to the one-to-two arrangement. But why do so many creatures have this same configuration?

Many scientists are unhappy with the "it's just history" explanation of homologous skeletal patterns. R.D.K. Thomas and W. E. Reif, for example, have developed an idea they call "the skeleton space." This sounds like a Halloween costume shop, but it's actually a cool way of saying that there are only a limited number of ways that geometric shapes and growing materials (like bones) can go together and still work well.

In their view, it's not "just history" or common descent that helps us understand why organisms have similar limbs. They claim that there are only a limited number of skeletal patterns because of the functional requirements of organisms. These are the limits imposed by geometry, and the characteristics of bones and the way they grow.**

So—was Darwin right when he said it was "hopeless" to try to understand vertebrate limb homology by referring to its function? Does the fact of form following function disprove the theory of common descent? Can homology be explained in more than one way? If so, which explanation best fits the facts? Time for a discussion! ■

* Francisco J. Ayala, "Evolution, The Theory of," *Encyclopedia Britannica*, 15th ed. (Chicago: Encyclopedia Britannica, 1988): 987.

** "The constraints of geometry, growth patterns, and raw materials… constitute formal causes of skeletal design. These do explain the convergence of numerous lineages on general patterns and the relatively complete exploitation of design elements defined in the skeleton space." R.D.K. Thomas and W. E. Reif, "The Skeleton Space: A Finite Set of Organic Designs," *Evolution*:47 (April 1993):353.

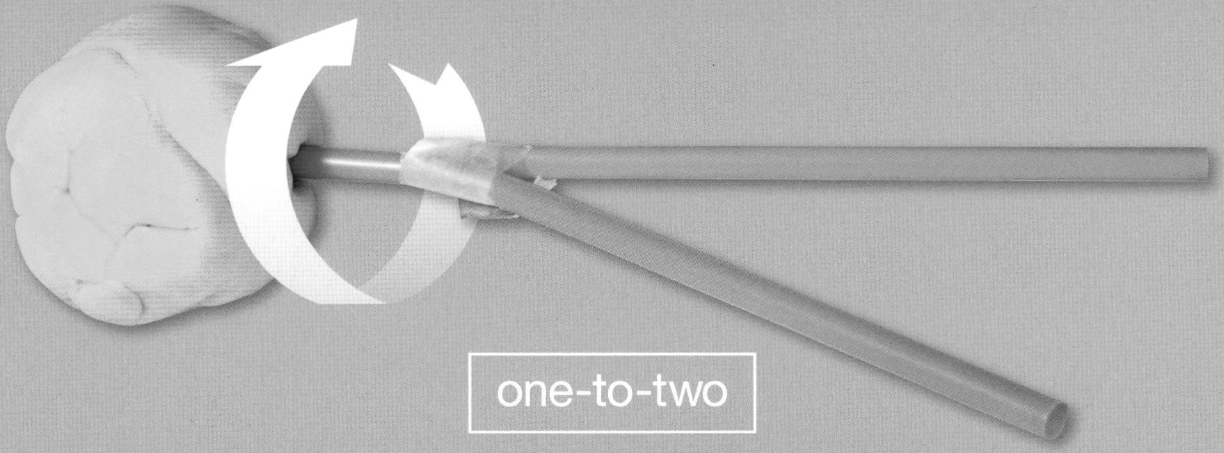

one-to-two

Neo-Darwinian biologists use the term "convergence" or "homoplasy" to describe similar structures that are not due to common ancestry but which are found in different types of organisms. They call these features convergent because they think that the evolutionary process has come together (converged) on the same structure two or more times in creatures that exist on very different branches of the Tree of Life.

What Does this Tell Us?

Convergence is a deeply intriguing mystery, given how complex some of the structures are. Some scientists are skeptical that an undirected process like natural selection and mutation would have stumbled upon the same complex structure many different times. Advocates of neo-Darwinism, on the other hand, think convergent structures simply show that natural selection can produce functional innovations more than once.

For other scientists, the phenomenon of convergence raises doubts about how significant homology really is as evidence for Common Descent. Convergence, by definition, affirms that similar structures do not necessarily point to common ancestry. Even neo-Darwinists acknowledge this. But if similar features can point to *having* a common ancestor—and to *not* having a common ancestor—how much does "homology" really tell us about the history of life?

ANATOMICAL HOMOLOGY: FURTHER DEBATE

Faced with the difficulties in explaining anatomical homology, some evolutionary biologists have given up on the notion. They argue that "the only way out of this dilemma is to stop talking about homology."[10] One such biologist, David Wake of the University of California—Berkeley, argues that "homology is not evidence of evolution, nor is it necessary to understand homology in order to accept or understand evolution."[11]

Other neo-Darwinian biologists argue that the reason for much of the misunderstanding is that homology at the genetic level cannot be equated with homology at the anatomical level. Biologist Gerhard Müller of the University of Vienna argues that "confusion arises because we are constantly mixing up levels, such as defining homologies of the anatomical level by shared developmental or genetic programmes."[12]

Critics argue that not being able to show a relationship between the levels of homology is not just a minor misunderstanding. It is a grave difficulty for the theory of Common Descent itself. They say that if Common Descent were true, then we would have every reason to expect that similar structures would arise from similar genes and developmental pathways. But quite often, they don't.

Some biologists suggest that the problems of understanding homology stem from Darwin himself, who re-defined homology as the result of common ancestry.[13]

This made the concept of homology circular, say many critics. If homology is defined as "similarity *due* to common descent," then to say that homology provides evidence *for* common descent is to reason in a circle. It would be like the political campaign slogan, "Governor Blinson is the kind of man who is right for the kind of times that demand a man like him." For this reason, Brian Goodwin argues that the Darwinian understanding of homology is ultimately empty. He stresses that biology must step beyond the classical Darwinian understanding of homology.

- Is the Darwinian definition of homology circular?
- Can the theory of Common Descent be reconciled with the new evidence from genetics and developmental biology?
- Are there other similarities that point to common ancestry?

As we will see in the next chapter, some biologists think there are other similarities that point strongly to common ancestry.

Endnotes

1 Charles Darwin, *On the Origin of Species* (Cambridge, Mass: Harvard University Press, 1964 [Facsimile of the First Edition, 1859]), Chapter 13: "Mutual Affinities of Organic Beings":434-439.

2 Alfred S. Romer, Thomas S. Parsons, *The Vertebrate Body.* (Philadelphia: W.B. Saunders, 1977):9.

3 Cuvier did not invoke a designer to explain these similarities, but he rejected common ancestry. He thought similar structures reflected functional requirements, not evolutionary history. He noted that all organisms have functional problems that must be solved, and that relatively few structures can solve them successfully. For Cuvier, it was not surprising that the same structures would be used in several different organisms.

see also E.S. Russell, *Form And Function: A Contribution to the History of Animal Morphology* (Chicago: University of Chicago Press, 1982):76. "Cuvier's view, on the contrary, is that the necessity of functional and ecological adaptation accounts for the repetition of the same types of structure. There are, of all the possible combinations of organs, only a few viable types—those whose structure is adapted to their life. Therefore it is reasonable that these few types should be repeated in innumerable exemplars.... He held that the resemblances between the organs of one class of animals and the organs of another were due to the similarity of their functions."

4 Goodwin contends that, "...just as liquids are states of matter which are organized in such a way that certain forms or morphologies are possible (spirals, waves, jets, etc., under various specific conditions), so organisms are states of matter organized in such a way that particular forms or morphologies are possible." "Is Biology An Historical Science?" in S. Rose and L. Appignanesi, eds., *Science and Beyond* (London: Basil Blackwell, 1986):57.

see also Brian Goodwin, *How the Leopard Changed Its Spots: The Evolution of Complexity* (New York: Charles Scribner's Sons, 1994).

5 Alfred S. Romer, Thomas S. Parsons, *The Vertebrate Body* (Philadelphia: W.B. Saunders, 1977):9-10.

6 David P. Mindell and Axel Meyer, "Homology evolving," *Trends in Ecology and Evolution* 16 (2001):343-440.

Claus Nielsen and Pedro Martinez, "Patterns of gene expression: homology or homocracy?" *Development, Genes, and Evolution* 213 (2003):149-154.

Jaume Baguñà and Jordi Garcia-Fernàndez, "Evo-devo: the long and winding road," *International Journal of Developmental Biology* 47 (2003):705-713.

Brian K. Hall, "Baupläne, phylotypic stages, and constraint: Why there are so few types of animals," *Evolutionary Biology* 29 (1996):215-253.

Lisa M. Nagy and Terri A. Williams, "Comparative limb development as a tool for understanding the evolutionary diversification of limbs in arthropods: challenging the modularity paradigm," in *The Character Concept in Evolutionary Biology*, ed. G. Wagner (New York: Academic Press, 2001):455-488.

7 Klaus Sander and Urs Schmidt-Ott, "Evo-Devo aspects of classical and molecular data in a historical perspective," *Journal of Experimental Zoology B* (Molecular and Developmental Evolution) 302 (2004):69-91.

8 Grace Panganiban and John L.R. Rubenstein, "Developmental functions of the Distal-less/Dlx homeobox genes," *Development* 129 (2002):4371-4386.

9 The full text reads, "Homology in some singular molecular components of eyes seems interesting but unsurprising; homology in complex genetic and developmental pathways for building eyes (as has now been discovered) was both unexpected under usual views of evolution and downright revisionary in forcing a rethinking of many previous certainties." Stephen Jay Gould, "Common pathways of illumination," *Natural History* 103 (Dec. 94):16.

10 David Wake, Homology - No. 222, CIBA Foundation Symposia Series, Novartis Foundation Symposium; Brian Hall (Dalhousie Univ., Halifax, Canada):45.

11 David Wake, op. cit., 27

12 Gerhard Müller, Homology - No. 222, CIBA Foundation Symposia Series, Novartis Foundation Symposium; Brian Hall (Dalhousie Univ., Halifax, Canada):44.

13 Developmental biologist Brian Goodwin has written extensively on this theme. As he puts it, "Darwin profoundly altered the meaning of homology when he argued that it should be understood in terms of descent from a common ancestor, replacing a logical [non-evolutionary] definition by an historical one." Brian Goodwin, "Is Biology an Historical Science?" S. Rose and L Appignanesi, eds. *Science and Beyond* (London: Basil Blackwell, 1986).

MOLECULAR HOMOLOGY

Molecular Homology: Case For

We have seen that some scientists—even some evolutionary biologists—have reservations about the Darwinian interpretation of anatomical homology. Nevertheless, many biologists say that another line of evidence, unknown in Darwin's time, provides decisive support for the theory of Universal Common Descent.

Specifically, they point to the striking similarity in molecules called proteins that all living things depend on for their very survival. In other words, molecules, as well as anatomical structures, display homology.

All cells are made up of proteins, the molecules that perform most of the important jobs in the cell. Proteins perform many tasks necessary for life.

Let's look at some protein tasks. For example, your body can't use sugar—at least, not in the form you eat it. Your body converts sugar into a form of energy it *can* use, and the conversion process begins with a protein called hexokinase. Skin is made largely of a different protein called collagen. Another protein enables you to see. When a photon of light reaches your retina, it first interacts with a protein called rhodopsin.

Proteins are extremely important because they build cellular structures, carry and deliver cellular materials, and act as enzyme catalysts in the chemical reactions that living things require. To perform these life-critical tasks, a typical cell contains thousands of different kinds of proteins, in the same way that a carpenter's workshop contains many different kinds of tools for various carpentry tasks.

Interestingly, proteins in all living things are made from the same basic "alphabet" of 20 amino acids.* Each type of protein is formed from a unique arrangement of these chemical "letters." The amino acid letters in proteins are arranged in precise sequences, like letters in an English sentence, organized and assembled according to coded instructions stored in another molecule known as DNA.

* Two more amino acids—selenocysteine and pyrrolysine—have recently been identified. This brings the total of known protein-forming amino acids to 22. This is so new that most biologists still refer to a 20-amino acid alphabet.

What does this have to do with the theory of Universal Common Descent? Lots. Biochemists have studied the amino acid sequences of proteins. Sometimes they compare the amino acid sequences for the same protein in different animals, and have discovered a number of cases where the sequences are remarkably similar. Is it possible that similar amino acid sequences show that these different animals originally had the same ancestor?

It turns out that molecular biologists can compare proteins from different species to see how different or similar they are. They have found that the same proteins in different organisms are similar in their "letter-by-letter" arrangement. For instance, the sequence of amino acids in hemoglobin—the protein that carries oxygen in the bloodstream—is very similar in chimpanzees and humans. Neo-Darwinists say the best explanation for this similarity is that chimpanzees and humans are descendents of a common ancestor—one that possessed an ancestral form of hemoglobin that later evolved in two slightly different ways.[1]

The same kind of sequence similarity we find in proteins is also found in DNA. DNA molecules perform a specific function in the cell: they carry the instructions for building proteins. These instructions are carried as a specific arrangement of chemical characters called "bases." There are four base characters: A (adenine), T (thymine), G (guanine), and C (cytosine).

Just as we saw with proteins, the sequence of these chemical characters—the "letters" in the genetic text—are arranged in very specific ways. Furthermore, the genes that code for the same proteins in different *{cont. on page 56}*

Remembering Key Terms

- Evolution #1: Change over time.
- Evolution #2: Universal Common Descent.
- Evolution #3: The Creative Power of Natural Selection.
- Monophyletic view: There is a single Tree of Life containing multiple branches.
- Polyphyletic view: There are separate Trees of Life, with some branching within each tree, but not between trees.

<<

DNA 101

You could think of DNA as looking like a twisted ladder. Genetic information is made up of varying combinations of the four base letters, and the genetic information is "spelled out" as you walk up the rungs of the ladder.

Each base letter on one side of the ladder is connected to a corresponding base letter on the other side. Connecting these "base pairs" forms a rung on the ladder.

A section of DNA text that carries the instructions for building a protein is called a "gene."

101

Those of you who have worked with "ham" radios know about that the heading above reads SOS, the international distress signal. In Morse Code, each letter of the alphabet has a corresponding visual pattern (dots and dashes) or sound pattern (*dits* and *dahs*).

The cell uses a code, too. Except, instead of using sequences of dots and dashes, the cell uses sequences of chemicals called "bases," stored in the molecule DNA. The four different bases have scientific names—adenine, thymine, guanine and cytosine—but many scientists often refer to them by their initials: A, T, G, and C.

When arranged properly, these chemical bases carry "assembly instructions" that tell the cell's internal machinery how to build proteins out of amino acids. Here's how it works. The genetic assembly instructions stored on a strand of DNA are reproduced on another molecule called "messenger RNA" (or mRNA). Messenger RNA also uses chemical bases to store genetic information, using a genetic alphabet similar to the one used by DNA. The one difference is that mRNA substitutes the base uracil—the letter U—for DNA's thymine (the letter T).

After it is formed, the mRNA travels to a molecular machine called a ribosome, which reads the mRNA assembly instructions. The instructions are made up of a series of genetic three-letter words called *codons*. Each codon tells the cell to attach a specific amino acid to a growing chain of other amino acids. For example, the mRNA word UUA tells the cell to attach the amino acid *leucine*, and AGA tells the cell to attach the amino acid *arginine*. There are also codons that tell the cell to start building the protein, and when to stop building it.

Scientists can actually chart the relationship between each individual codon and a specific amino acid. Biologists refer to the whole set of these relationships as "the genetic code." (You can see a chart of the standard genetic code on the page facing).

Now, what does this have to do with the theory of Universal Common Descent? Quite a lot, say advocates of the theory. They have long argued that nearly all organisms use the same codon-to-amino-acid code. This is called the *universality* of the genetic code. They say this evidence points to a universal common ancestor.

Why does a universal code imply a universal common ancestor? Here's what advocates of Common Descent say:

Once a code is established, it can't change without destroying the organisms that depend on it. Any small change in the codon-amino acid assignments means that the right amino acids will not be added in the right place. The resulting chains of amino acids might not fold properly, and might not function as a protein. This could be disastrous for the organism.

To see the effects of changing a code, imagine typing a history paper on a school computer. What you don't know is that someone has secretly changed the relationships between the keys on the keyboard and the letters that appear on your screen. For example, when you hit the key marked "a," the letter "x" appears. Worse yet, every time you hit a "period," the letter "e" appears. Now, imagine turning that paper in without giving an explanation about the change in the keyboard code. Will the paper make sense? What kind of grade do you think you'll get?

Advocates of Common Descent recognize that organisms are also sensitive to any change in the codon-to-amino-acid assignments. Instead of producing a bizarre sentence, the cell affected by a random change in code would produce garbled sequences of amino acids that won't fold into functioning proteins, which could literally be a question of life or death.

Now, think about what that means. If the codon-amino acid assignments can't change without destroying the cell's critical proteins, then once life has originated, all subsequent forms of life must evolve with that same original code or die. Thus, the presence of a single common code today would seem to point back to a single common ancestor—the first form of life that arose with life's first functioning, and now unchangeable, genetic code. If all forms of life evolved from a single common ancestor, then we should expect to find the same code in every form of life today. For this reason many biologists see the universality of the code as prime evidence for Universal Common Descent. ■

	U	C	A	G
U	UUU Phenylalanine UUC Phenylalanine UUA Leucine UUG Leucine	UCU Serine UCC Serine UCA Serine UCG Serine	UAU Tyrosine UAC Tyrosine UAA Stop UAG Stop	UGU Cysteine UGC Cysteine UGA Stop UGG Tryptophan
C	CUU Leucine CUC Leucine CUA Leucine CUG Leucine	CCU Proline CCC Proline CCA Proline CCG Proline	CAU Histidine CAC Histidine CAA Glutamine CAG Glutamine	CGU Arginine CGC Arginine CGA Arginine CGG Arginine
A	AUU Isoleucine AUC Isoleucine AUA Isoleucine AUG Methionine (Start)	ACU Threonine ACC Threonine ACA Threonine ACG Threonine	AAU Asparagine AAC Asparagine AAA Lysine AAG Lysine	AGU Serine AGC Serine AGA Arginine AGG Arginine
G	GUU Valine GUC Valine GUA Valine GUG Valine	GCU Alanine GCC Alanine GCA Alanine GCG Alanine	GAU Aspartic GAC Aspartic GAA Glutamic GAG Glutamic	GGU Glycine GGC Glycine GGA Glycine GGG Glycine

The codons in this chart are those that are present on messenger RNA, which transmits a copy of the genetic information from the original DNA strand.

Why are some codons printed in blue? If you want to find out, check out the sidebar on page 58.

"The greater the differences in the DNA,...the longer the time since two organisms shared a common ancestor. This DNA evidence for evolution has confirmed evolutionary relationships derived from other observations."

U.S. National Academy of Sciences, Teaching About Evolution and the Nature of Science *(1998).*

organisms have remarkably similar sequences. This is just what we would expect if these genes and proteins originated from a common ancestor.

Evolutionary biologists see two other patterns that support this view. First, they note that, "the code used to translate [DNA] sequences into amino acid sequences is essentially the same in all living organisms." Thus, as the United States National Academy of Sciences contends, "this unity of composition and function is a powerful argument in favor of... common descent."[2]

Second, according to many neo-Darwinists, the family trees constructed from molecular homologies match the family trees constructed from anatomical homologies. The U.S. National Academy of Sciences recently published a booklet explaining the current evidence for evolutionary theory. The booklet says that family trees constructed from the molecules myoglobin and hemoglobin, "agreed completely with observations derived from paleontology and anatomy about the common descent of the corresponding organisms." Thus, they conclude, "The evidence for evolution from molecular biology is overwhelming and is growing quickly."[3]

Molecular Clock

According to many evolutionary biologists, the evidence from molecular homology points to a single Tree of Life. But they also argue that the evidence tells us *when* the branches of the Tree of Life split off. In other words, it tells us the last point at which two (or more) species shared a common ancestor. The reasoning looks like this: as proteins evolve over time, their amino acid sequences change a little bit at a time. The more time they have had to change, the more different they appear from one another. By comparing the differences between the sequences of the *same* protein in *different* species, evolutionary biologists can determine when the two species split off from their common ancestor and began to evolve independently. In this way, the differences in sequences function like a clock—a "molecular clock." As the National Academy of Sciences has explained, "The concept of a molecular clock is useful for two purposes. It determines evolutionary relationships among organisms, and it indicates the time in the past when species started to diverge from one another." [4] This evidence, says the National Academy, is further support for Common Descent (Evolution #2).

Molecular Homology: Reply

Critics of the argument from molecular homology agree that the molecules in living things exhibit many remarkable similarities in sequence. They interpret this evidence differently, however. Critics argue that similarities may reflect common functional requirements, rather than a common evolutionary past. And they point out that some molecular evidence challenges Common Descent.

Critics acknowledge, for example, that hemoglobin molecules in different organisms are similar in sequence and in structure. But why should this be surprising, they argue, when these molecules perform the same function: binding and releasing oxygen? As with anatomical similarity, the evidence of sequence similarity admits more than one explanation. Molecular similarities may result from common ancestry or they may reflect common functional requirements.

Problems with the Family Tree?

But some critics go further. They argue that other molecular evidence cannot be reconciled with the theory of Universal Common Descent. For example, if Darwin's single Tree of Life is accurate, then we should expect that different types of biological evidence would all point to that same tree. A "family history" of organisms based on their anatomy should match the "family history" based on their molecules (such as DNA and proteins).[5]

Many scientists have argued that, contrary to the claims of the U.S. National Academy of Sciences booklet, this is frequently not the case. Evolutionary biologist Michael Lynch has noted that creating a clear picture of evolutionary relationships is "an elusive problem." He also notes that "analyses based on different genes—and even different analyses based on the same genes—[yield] a diversity of phylogenetic trees."[6]

A "family tree" based on anatomy may show one pattern of relationships, while a tree based on DNA or RNA may show quite another.[7] For example, one analysis of the mitochondrial cytochrome *b* gene produced a "family tree" in which cats and whales wound up in the order Primates.[8] Yet, an anatomical analysis says that cats belong to the order Carnivora, while whales belong to Cetacea—and neither of them are Primates.

In fact, a family tree based on one protein may differ from a family tree based on a different protein.[9] Sometimes, when two different laboratories analyze the same protein, they produce different family trees.[10]

Universal No More

Many organisms use genetic codes that differ from the so-called "universal" code. Remember the codons printed in blue in the chart on page 55? In some organisms these codons send a different signal than they do in the standard code.

For example:

- In single-celled eukaryotes such as *Tetrahymena*, stop codons (e.g., UAG and UAA) have been assigned to the amino acid glutamine.
- In the fungus (yeast cell) *Candida*, the codon for the amino acid leucine, CUG, now codes for a different amino acid, serine.
- In the bacterium *Micrococcus*, the codon AGA (for arginine) has disappeared from the organism's genome.

See Manuel A.S. Santos and Mick F. Tuite, "Extant Variations in the Genetic Code," in *The Genetic Code and the Origin of Life*, ed. L. Ribas de Pouplana (New York: Kluwer Academic, 2004):183-200.

<<

Yet, if all these organisms really did evolve from a single common ancestor, only *one* of the trees can be right. Critics point out that the real problem may be that Universal Common Descent is wrong. In other words, maybe the reason the family trees don't agree is that the organisms in question never did share a common ancestor.

Even some evolutionary biologists agree. Carl Woese of the University of Illinois, for instance, now thinks that biology must abandon what he calls Darwin's "Doctrine of Common Descent." Based on his study of the different domains of life, Woese says life probably had multiple, independent starting points.[11]

Genetic Code Not Universal

Another widely used argument for Universal Common Descent has recently come under fire. Biologists have long thought that the genetic code is basically the same in all living organisms—that is, genes "code for" the same protein in almost identical ways in almost all living things. As our SOS discussion illustrated *(See pages 54-55)*, it is difficult to see how the codon-amino acid assignments could change without killing the host organism. That's why evolutionary biologists have argued that the code we have today is the same as the code in the first living organism and why a universal genetic code points to a universal common ancestor.

But *is* the genetic code universal? It turns out that it's not.[12] Since 1985 molecular biologists have discovered at least 18 different genetic codes in various species.[13] Many of these are significantly different from the standard code.[14]

For example, the standard code has three different mRNA stop codons: UGA, UAA, and UAG. (A "stop codon" tells the cell to stop building—the protein is complete.) However, some variant codes have only one stop codon, UGA. The other "universal" stop codons now code for the amino acid glutamine.

It's very hard to see how an organism could have survived a transformation from the standard code to this one. Changing to this new code would cause the cell to produce useless strings of extra amino acids when it should have stopped protein production. You might be tempted to think that this isn't a problem, because more is better, right? Not always. Having extra amino acids in a critical protein is not like having extended scenes on a DVD movie. The protein's ability to function depends on its final shape, which depends on how it folds. Some amino acid sequences are robust—able to function in spite of mutational alteration. In other cases, extra amino acids will cause the chain of amino acids to fold differently and lose function.

With a non-robust sequence in a critical protein, any change in the genetic code leaves the organism with essentially two choices: 1) Die. 2) Simultaneously acquire a new set of genetic information that matches the new code. Nearly all biologists think that the second scenario is way too improbable.

For this reason, advocates of universal common descent have always assumed that a change in code was not survivable, and that once the code evolved, it couldn't evolve any further. They argued that a single common genetic code implied a single origin of life—a universal common ancestor.

By that same logic, if a single code implies a single origin of life, do multiple codes imply multiple separate origins? Some scientists think this is a possibility, saying that the evidence may point to a polyphyletic view of the history of life.[15]

Molecular Clock Reply

Critics also dispute both the accuracy and the importance of the "molecular clock." They dispute the accuracy because of many known problems with calibrating such clocks.[16]

To time something accurately, you must know that your watch runs at a constant rate—that it doesn't speed up or slow down. Unfortunately, say the critics, the rate of mutation varies in response to a number of environmental factors.[17] As a result, even if we knew when species diverged, we couldn't be sure that the molecular clock was "ticking" at a constant rate.[18]

Additionally, and far more importantly, critics question the logic of molecular clock theories, and claim that its relevance as "evidence" for Common Descent is dubious at best. Here's why.

Molecular clock calculations tell how long ago a common ancestor for two or more organisms might have existed—*if* one takes for granted that such an organism *must* have existed. But critics say this takes for granted the very point it is supposed to be proving. Logically, you can't use a method that *assumes* the existence of a common ancestor to *prove* the existence of a common ancestor. Yet, this is precisely what the molecular clock does. Critics say this is circular reasoning—a logical fallacy called "begging the question."

A New Puzzle

But Darwin's critics say the biggest challenge to molecular homology is the puzzle now emerging from the cutting-edge scientific discipline of genome sequencing. Genome sequencing identifies the location of all the protein-building instructions (genes) contained in the DNA. Genomic researchers can now map out the entire DNA sequence of a species. How do they do this?

Genes that code for proteins have distinctive properties. Remarkably, they exhibit many of the same characteristics as English grammar. Most English sentences begin with a capital letter and end with a period.* The capital letter indicates the start of the sentence, the period indicates the end, and there are strings of meaningful characters in between.

Genes behave in much the same way, except that instead of a capital letter, a gene uses a "start site"—a short section of code that tells the "copy machine" (called a ribosome[19]) where to set up shop. Instead of a period, the gene has a "stop site"—a short section of code that tells the ribosome when to stop construction.[20] If the ribosome starts where it should, follows the instructions, and stops when it should, it produces the right protein.

---

* No, of course we haven't forgotten about question marks and exclamation points. Just work with us, OK?

Here's the point. Scientists can recognize genetic "sentences" because they exhibit the proper genetic "grammar" (start sites, stop sites, and ribosomal binding sites).

Computer programs analyze long strings of DNA sequences, and can recognize the distinctive start-and-stop regions. They can then identify sequences that might code for proteins—in other words, sequences that might be genes. This is rather like detecting a meaningful phrase within a string of letters.

*jzvrb#mk*ldybc%ct**O**rder*
cheeseandpepperonipizza.
wpsl$jakzx^rgveioybr+n&

Molecular biologists have analyzed millions of base pairs of DNA from various organisms. They have been surprised to learn that a large number of genes code for proteins whose function we don't understand yet. They call these ORFan genes.* Unlike "order cheese and pepperoni pizza" in our example, the meaning of ORFan genes is unknown.[21] Biologists can tell that ORFans are genes, but they don't yet know what these genes do. (That's how they got the name "genes of unknown function.") But the real puzzle is their mysterious origin.[22]

According to evolutionary theory, new genes arise from old genes by mutation. This process should leave a trail of evidence behind—clues that would allow us to figure out the ancestry of the genes. New genes should resemble the older "ancestor" genes. However, these newly discovered genes of unknown function do not match any other sequence that codes for a known protein.[23] As molecular biologist Russell Doolittle notes, "Where these unique sequences are coming from and what they do remain baffling mysteries."[24] Is it possible that these puzzling genes arose independently in various groups of organisms?

Some scientists think so. As genomic researchers Peterson and Fraser have speculated, "[another] extremely interesting possibility is that many gene functions have evolved independently more than once since the beginning of cellular life on the planet."[25]

* The genes have been dubbed "ORFans"—pronounced "orphans" ("ORF" stands for "Open Reading Frame," the technical term for a DNA sequence that codes for a protein).

MOLECULAR HOMOLOGY: FURTHER DEBATE

Do the puzzles of conflicting phylogenetic trees or ORFans (genes of unknown origin and function) undermine the theory of Universal Common Descent? Most evolutionary biologists say no. These scientists say that molecular biology simply needs new tools to understand these puzzles.

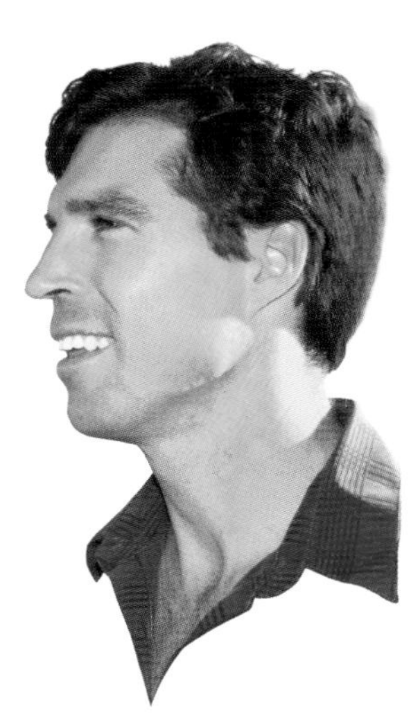

Others are not so sure. Molecular evolutionist Michael Syvanen of the University of California-Davis argues that, "there is no reason to postulate that a LUCA (Last Universal Common Ancestor) ever existed."[26] Biologist Malcolm Gordon of UCLA argues that the single branching-tree picture of life's history is not accurate, but that life must have had multiple, independent starting points.[27]

Faced with the discovery of multiple genetic codes, some scientists are reconsidering the question of whether there are evolutionary mechanisms that could cause the code to change after all. Some of these scientists are convinced there are.[28] For this reason, they doubt that discovery of multiple codes actually challenges Common Descent.

Nevertheless, research continues...and so does this debate.*

* For the latest information on this debate, check out **www.exploreevolution.com**

Endnotes

1 Maximilian J. Telford and Graham E. Budd, "The place of phylogeny and cladistics in Evo-Devo research," *International Journal of Developmental Biology* 47 (2003):479-490.

2 *"Science and Creationism: A View from the National Academy of Sciences,"* Second Edition. Washington, D.C.: National Academy Press:17.

3 *"Science and Creationism: A View from the National Academy of Sciences,"* op. cit., p. 18.

4 *"Science and Creationism: A View from the National Academy of Sciences,"* op. cit., p. 19.

5 Luciano Brocchieri, "Phylogenetic inferences from molecular sequences: review and critique," *Theoretical Population Biology* 59 (2001):27-40.

6 Michael Lynch, "The Age and Relationships of the Major Animal Phyla." *Evolution* 53 (1999):319-325.

7 Ying Cao, Peter J. Waddell, Norhiro Okada, and Masami Hasegawa, "The complete mitochondrial DNA sequence of the shark Mustelus manazo: Evaluating rooting contradictions to living bony vertebrates," *Molecular Biology and Evolution* 15 (1998):1637-1646.

Christiane Delarbre, Ann-Sofie Rasmussen, Ulfur Arnason, and Gabriel Gachelin, "The complete mitochondrial genome of the hagfish Myxine glutinosa: unique features of the control region," *Journal of Molecular Evolution* 53 (2001):634-641.

Melvin R. Duvall and Autumn Bricker Ervin, "18S gene trees are positively misleading for monocot/dicot phylogenetics," *Molecular Phylogenetics and Evolution* 30 (2004):97-106.

S. Blair Hedges and Charles G. Sibley, "Molecules vs. morphology in avian evolution: The case of the 'pelecaniform' birds," *Proceedings of the National Academy of Sciences USA* 91 (October 1994):9861-9865.

Ann-Sofie Rasmussen and Ulfur Arnason, "Molecular studies suggest that cartilaginous fishes have a terminal position in the piscine tree," *Proceedings of the National Academy of Sciences USA* 96 (1999): 2177-2182.

A. Rokas, N. King, J. Finnerty, and S.B. Carroll, "Conflicting phylogenetic signals at the base of the metazoan tree," *Evolution and Development* 5 (2003):346-359.

Yuri I. Wolf, Igor B. Rogozin, and Eugene V. Koonin, "Coelomata and not ecdysozoa: evidence from genome-wide phylogenetic analysis," *Genome Research* 14 (2004):29-36.

8 Michael S.Y. Lee, "Molecular phylogenies become functional," *Trends in Ecology and Evolution* 14 (1999):177-178.

9 A "family tree" based on the molecule 18s RNA can be different from one based on a different molecule 28s RNA, as reported by R. Christen, et al., *EMBO Journal* 10 (1991):499-503.

10 The same molecule (18s RNA) produced one "family tree" when analyzed in one lab (Birgitta Winnepenninckx, et al., *Molecular Biology & Evolution* 12 (1995):641-649 and 1132-1137) and produced a different "family tree" when analyzed in a different lab (Anna Marie A. Aguinaldo & James A. Lake, *American Zoologist* (1998).

11 Carl Woese, "On the evolution of cells," *Proceedings of the National Academy of Sciences* 99 (2002):8742- 8747. "Extant life on Earth is descended not from one, but from three distinctly different cell types."

12 Philip Cohen, "Renegade code," *New Scientist* 179 (2003):34-38.

13 As of April 2005, the most recent list of genetic codes could be found at http://www.ncbi.nlm.nih.gov/Taxonomy/Utils/wprintgc.cgi?mode=c

If it's not there, go to http://www.ncbi.nlm.nih.gov/entrez/ and do a search for the term "genetic codes."

14 Lluís Ribas de Pouplana, ed., *The Genetic Code and the Origin of Life* (New York: Kluwer Academic/Plenum Publishers, 2004).

see also Christine Fenske, Gottfried J. Palm, and Winfried Hinrichs, "How unique is the genetic code?" *Angewandte Chemie International Edition* 42 (2003):606-610.

15 "The pervasiveness of the standard genetic code has often been cited as evidence of a single origin of life event (Mayr 1982). The other side of the coin is that the existence of several different genetic codes is evidence of a number of independent origin of life events." Hubert Yockey, *Information Theory and Molecular Biology* (Cambridge, England: Cambridge University Press, 1992):202-203.

16 James W. Valentine, David Jablonski, and Douglas H. Erwin, "Fossils, molecules and embryos: new perspectives on the Cambrian explosion," *Development* 126 (1999):851-859.

17 According to James W. Valentine, David Jablonski, and Douglas Erwin, these environmental factors might include the collapse of magnetic fields, and mass extinctions (which may create environmental niches).

18 One way to solve this problem would be to cross-reference the "dates of divergence" we get from the molecular clock with the "first appearances" we find in the fossil record. Unfortunately, scientists have encountered a practical problem. As Michael Lee, a molecular evolutionist at the University of Queensland in Australia noted, actual fossil dates are often unreliable—making any molecular clock set by them unreliable as well. Michael S.Y. Lee "Molecular clock calibrations and metazoan divergence dates" *Journal of Molecular Evolution* 49 (1999):385-391.

see also Dan Graur and William Martin, "Reading the entrails of chickens: molecular timescales of evolution and the illusion of precision," *Trends in Genetics* 20 (February 2004):80-86. "In this article, we document the manner in which a calibration point that is both inaccurate and inexact—and in many instances inapplicable and irrelevant—has been used to produce an exhaustive evolutionary timeline that is enticing but totally imaginary."

19 The "copy machine" is actually a combination of proteins and a skeleton of catalytic RNAs.

20 There are even rules of genetic grammar. The Open Reading Frame is a set of instructions telling the ribosome to bind at the start site and transcribe until it finds a stop site.

21 Daniel Fischer and David Eisenberg, "Finding Families For Genomic ORFans," *Bioinformatics* 15 (1999):759-762.

22 Naomi Siew and Daniel Fischer, "Analysis of Singleton ORFans in fully sequenced microbial genomes," *Proteins: Structure, Function, and Genetics* 53 (2003):241-251.

23 "If proteins in different organisms have descended from common ancestral proteins by duplication and adaptative variation, why is it that so many today show no similarity to each other? Why is it that we do not find today any of the necessary 'intermediate sequences' that must have given rise to these ORFans?" Naomi Siew and Daniel Fischer, "Twenty thousand ORFan microbial protein families for the biologist?" *Structure* 11 (Jan 2003):7-9.

24 Russell F. Doolittle, "Microbial genomes multiply," *Nature* 416 (18 April 2002):698.

25 Scott N. Peterson and Clare Fraser "The complexity of simplicity." *Genome Biology* 2 (2003):1-8.

26 Michael Syvanen, "On the occurrence of horizontal gene transfer among an arbitrarily chosen group of 26 genes," *Journal of Molecular Evolution* 54:265.

Molecular evolutionist W. Ford Doolittle, of Dalhousie University, suggests an even more radical idea. He argues that the concept of molecular homology itself must be revised. “Homology is still a funny word,” Doolittle writes. “In the context of proteins and genes, it makes sense only if we don’t think about it too deeply” (emphasis added). He says the standard understanding of homology is no longer relevant, and is becoming “… a useless word unless we redefine it to mean something like ‘statistically more similarity than we would expect on the basis of chance.’” He concludes, “It is ironic that the words we seem to need in order to think productively about biology, words such as ‘homology’, ‘individual’, ‘organism’, and ‘species’ have no precise meaning.” W. Ford Doolittle, “The nature of the universal ancestor and the evolution of the proteome,” *Current Opinion in Structural Biology* 10 (2000):357-358.

27 Malcolm S. Gordon, “The concept of monophyly: a speculative essay,” *Biology and Philosophy* 14 (1999):331-348.

28 Manuel A.S. Santos and Mick F. Tuite, “Extant Variations in the Genetic Code” in *The Genetic Code and the Origin of Life* ed. Lluis Ribas de Pouplana (New York: Kluwer Academic/Plenum Publishers, 2004):183-200.

EMBRYOLOGY >

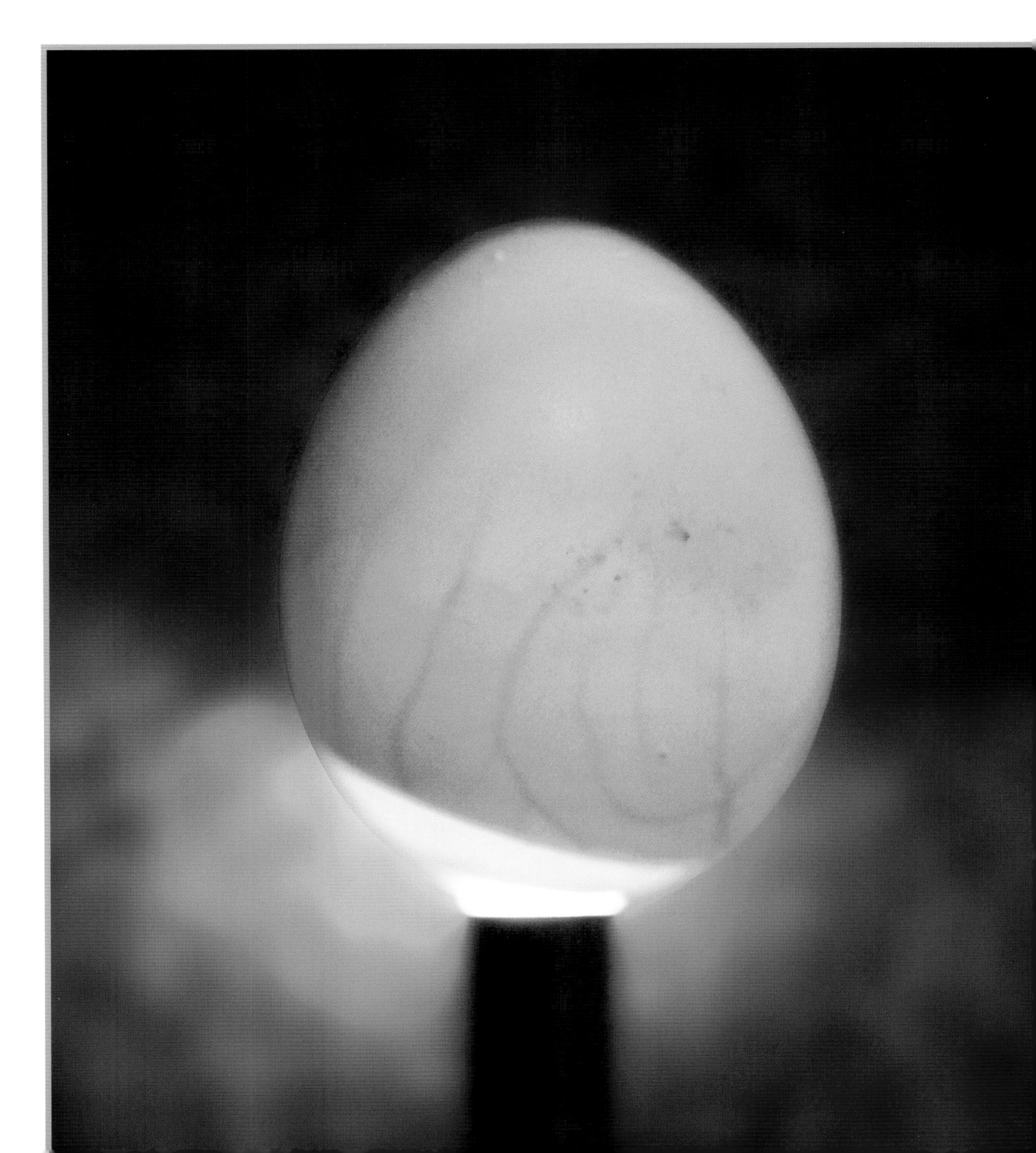

EMBRYOLOGY: CASE FOR

Embryology is the study of the development and formation of embryos—plants or animals in the earliest stages of life. One of Darwin's classic arguments for Common Descent is based on the similarity of embryos of different organisms.

Writing to his close friend, botanist Joseph Hooker, Darwin said, "Embryology is my pet bit in my book [the *Origin*]." In fact, Darwin thought that the study of embryological development provided "by far the strongest single class of facts in favor of" his theory of Common Descent.*

Darwin noticed certain similarities in the embryos of vertebrate animals, similarities he thought were especially great during the embryo's earliest stages of development. As Darwin put it, "Embryos of different species within the same class generally, but not universally, resembl[e] each other." [1] In his view, this observation established two things.

First, it established that organisms had descended from a common ancestor. Why did he think that? Darwin thought that the embryos of modern vertebrates are similar because they evolved from an ancestral form that had many, but not all, of the embryonic features of modern vertebrates.

Second, Darwin thought that the observable similarities in different embryos revealed what the ancestors to these organisms would have looked like. The embryo, argued Darwin, "is the animal in its less modified state; and... it reveals the structure of its progenitor." [2]

German embryologist Ernst Haeckel extended and popularized Darwin's two main ideas about embryology. Following Darwin's lead, Haeckel tried to discover the evolutionary history of various animals by studying their embryos. He produced a set of influential drawings showing that the embryos of various classes of vertebrates were very similar during their earliest stages of development. *(See Figure 4:1)* He also formulated and popularized his famous "Biogenetic Law," which states "ontogeny recapitulates phylogeny." In everyday language, that means the embryo's step-by-step process of development (ontogeny) repeats (recapitulates) the evolutionary history of the species (phylogeny).

* Embryological development is the step-by-step construction of animals, beginning with the fertilized egg and ending with the adult organism.

Modern evolutionary biologists have modified Darwin's and Haeckel's ideas. While most neo-Darwinists still accept that early embryological similarities point to common descent, they no longer think that embryos reveal the *adult* form of their evolutionary ancestors. Rather, some biologists, including Nobel laureate Peter Medawar, now think that embryos tell us what the *embryos* of their evolutionary ancestors might have looked like. He notes that embryos show increasing divergence from a common embryonic state, and explains, "The embryos and young of related animals resemble each other more closely than the adults into which they develop."[3] Thus, he concludes that there is an element of truth to the so-called law of recapitulation. A number of evolutionary biologists now share this view.

In any case, many evolutionary biologists continue to affirm Darwin's central claim about embryology: the patterns of embryonic similarity in animals can best be explained by descent from a common ancestor.

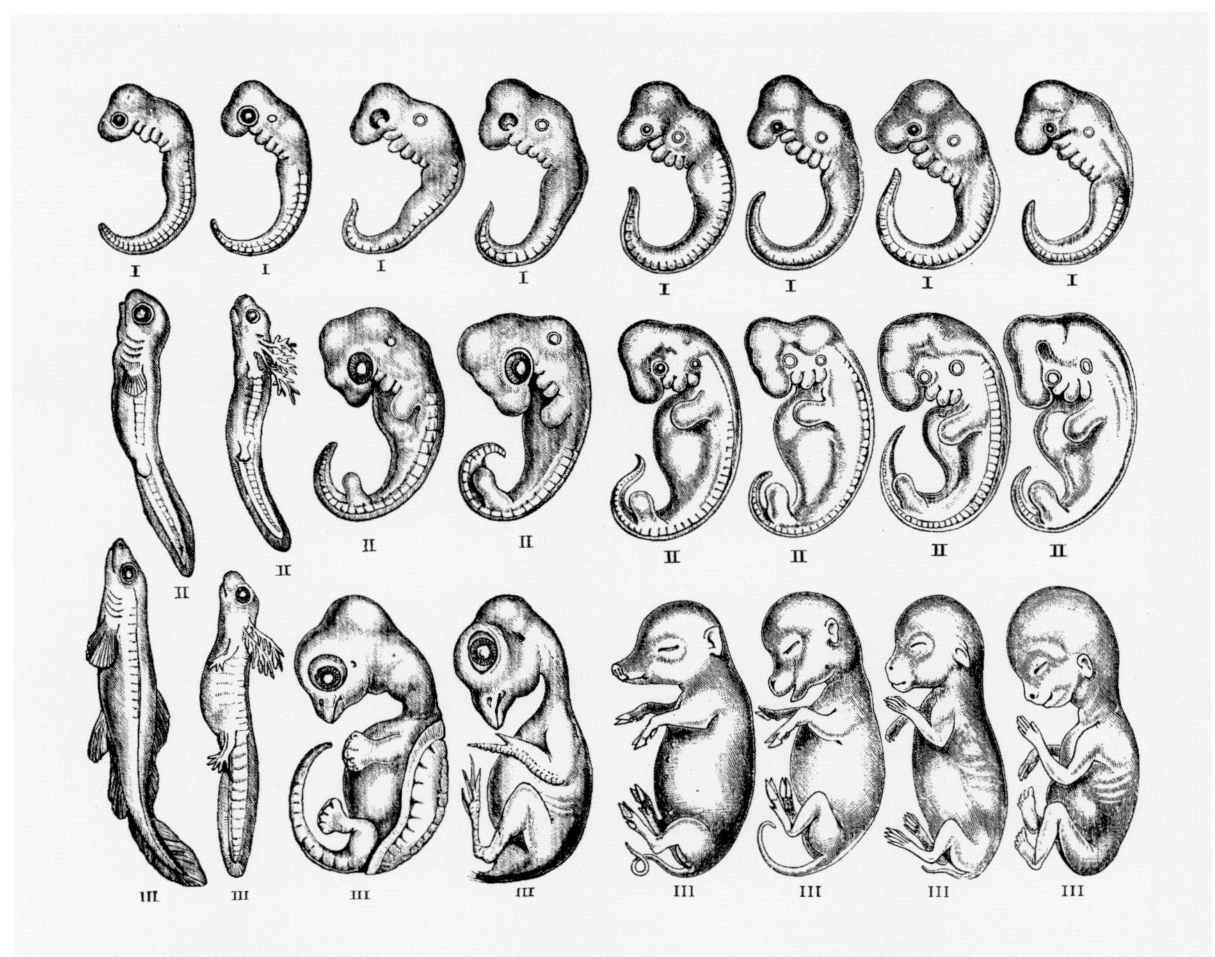

Figure 4:1 Haeckel's Embryos

Embryology: Reply

*Critics of the argument from embryology agree that common descent might be a reasonable inference to draw from the similarity of embryos—if embryos really **were** similar in their earliest stages of development. But they're not, say most embryologists. Curiously, scientists have known this since the 19th century.*

In 1894, Adam Sedgwick, an embryologist at Cambridge University, challenged Darwin's two claims: 1) that vertebrate embryos were more alike than the vertebrate adults, and 2) that the younger the embryos, the greater the resemblance. This was "clearly not" the case, Sedgwick noted.

Even the embryos of "closely allied animals," such as chickens and ducks, display specific differences very early in development. "I can distinguish a [chicken] and a duck embryo on the second day," he wrote. Comparing the embryos of a chicken and a shark, he continued, "There is no stage of development in which the unaided eye would fail to distinguish them with ease." He added that these early differences come as no surprise, saying, "Every embryologist knows that they exist and could bring forward innumerable instances of them."[4]

This raises an interesting question. If, in fact, the early embryos of different species show marked differences, why did the erroneous claim of similarity become so widespread? Critics of this argument cite two reasons. The first was the widespread influence of Ernst Haeckel's drawings, illustrating what he called the "pharyngula" stage of various vertebrate embryos. It turns out that Haeckel's drawings misrepresented the features of the embryos, exaggerating their apparent similarities to support the argument for Common Descent.

The second reason is that both Darwin's and Haeckel's comparisons left out the earliest stages of development.* This omission is critical, because vertebrate embryos differ very strikingly at this stage. The stage Haeckel labeled the "first" is actually *midway* through development. Embryologists such as Sedgwick knew this, but their publications challenging the Darwinian interpretation were lost beneath the overwhelming popularity of Haeckel's inaccurate (and possibly fraudulent) drawings. These drawings soon took on a life of their own, moving from one

---

*Whether this omission was intentional or not is a matter of some debate.

biology textbook to another, accompanied by the erroneous claim that vertebrate embryos are most similar in their earliest stages. This error even crept into the *Encyclopedia Britannica*, and remains in many modern high school and college biology textbooks.[5]

In 1997, an international team of scientists, led by the embryologist Michael Richardson, compared Haeckel's drawings to photographs of actual embryos at various developmental stages. They found that Haeckel had distorted the evidence at every turn, leading Richardson to tell the journal *Science* that "it looks like it's turning out to be one of the most famous fakes in biology."[6] In response, some biology textbooks are finally beginning to drop the erroneous drawings. "We do, I think, have the right," wrote Stephen Jay Gould, "to be both astonished and ashamed by the century of mindless recycling that has led to the persistence of these drawings in a large number, if not a majority, of modern textbooks."[7]

So, the pictures were fabricated and the facts were distorted. Anyone can verify this by looking at the recent scientific literature, say the critics *(see Figure 4:2)*. Critics of the embryology argument remind us that if these vertebrates shared a common ancestor, then these "descendant" embryos should be most similar to each other in the earliest stages of their development. But they aren't.

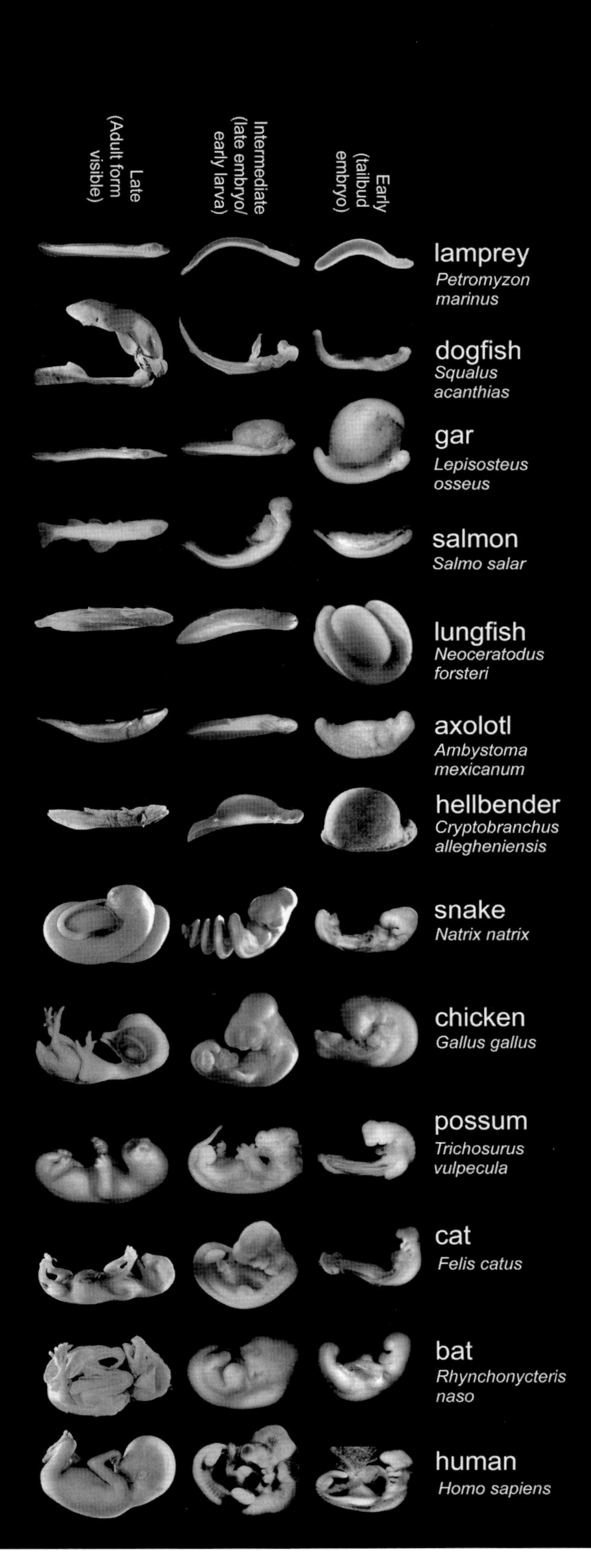

Figure 4:2 Photographs of the stages of real embryonic development. *We thank Michael Richardson for permission to use figure from this publication: Michael K. Richardson et al., "There is no highly conserved embryonic stage in the vertebrates: implications for current theories of evolution and development," Anatomy and Embryology 196 (1997):91-106.*

EMBRYOLOGY: FURTHER DEBATE

There is a wide range of opinions on the merits of the case from embryology. According to some critics of the standard embryology argument, the evidence strongly suggests that biologists should re-evaluate whether all animals shared a common ancestor.[8]

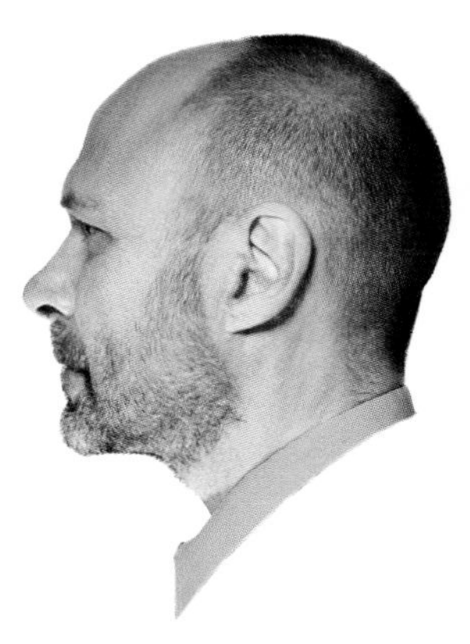

Other scientists defend the standard evolutionary interpretation of embryology. Alan D. Gishlick, writing for the National Center for Science Education, says "[I]t really does not matter what Haeckel thought or whether his drawings are accurate: modern comparative embryology does not stand or fall on the accuracy of Haeckel...."[9] Anthropologist Eugenie Scott, also of the National Center for Science Education, defends the drawings as rightly belonging among the "well-known and frequently repeated examples of principles or mechanisms of evolution... Textbooks use them because they communicate these basics clearly to uninformed students."[10]

Dissimilarity or Adaptation?

Still other scientists say that in some cases, early embryos appear different because those stages have been modified to adapt to environmental conditions. Darwin was aware of these dissimilarities, but he argued that they did not disprove Common Descent. As Darwin explained, some groups of embryos have been "so greatly modified [by adaptations]* as no longer to be recognized."[11]

Critics of the embryology argument are troubled by the logic of this claim. By arguing that common descent predicts both embryonic similarity *and* dissimilarity, Darwinists have effectively made it impossible to challenge the theory with counterevidence. When the case is stated this way, common descent would be consistent with *whatever* we observe in embryos. Critics say it's like predicting, "In a coin toss, the result will be heads or tails." The prediction yields no useful information, despite being accurate 100% of the time. In the same way, when predictions of embryonic similarity *and* dissimilarity are fulfilled, have we learned anything meaningful? Critics say no. Neo-Darwinists disagree. They say that a combination of similarities and dissimilarities is just what one would expect if organisms had, in fact, *evolved* (descended with modifications) from a common ancestor.

No doubt the debate will continue.

Endnotes

1 Charles Darwin, *On the Origin of Species* (Cambridge, Mass: Harvard University Press, 1964 [Facsimile of the First Edition, 1859]), Chapter 13: "Mutual Affinities of Organic Beings":442.

2 Charles Darwin, *op.cit.*:449.

3 Peter Medawar, Nobel Conference XVIII, *Darwin's Legacy*, (San Francisco, CA: Harper & Row, 1983):50.

4 Adam Sedgwick; "On the law of development commonly known as von Baer's Law; and on the significance of ancestral rudiments in embryonic development," *Quarterly Journal of Microscopical Science* 36 (1894):35-52.

5 Peter H. Raven & George B. Johnson, *Biology* (New York: McGraw Hill, 6th edition, 2002):1229.

Douglas J. Futuyma, *Evolutionary Biology* (Sunderland, MA: Sinauer Associates, Inc, 3rd edition, 1998):653

6 Elizabeth Pennisi, "Haeckel's Embryos: Fraud Rediscovered," *Science* 277 (1997):1435.

see also Michael K. Richardson, James Hanken, Mayoni L. Goonеratne, Claude Pieau, Albert Raynaud, Lynne Selwood, and Glenda M. Wright, "There is no highly conserved embryonic stage in the vertebrates: implications for current theories of evolution and development." *Anatomy and Embryology* 196 (1997):91-106.

7 Stephen Jay Gould, "Abscheulich! Atrocious!" *Natural History* (March, 2000):42-49.

8 Jonathan Wells, *Icons of Evolution: Science or Myth?* (Washington, D.C., Regnery Press, 2000):81-109.

9 Alan D. Gishlick, "Icons of Evolution?: Why much of what Jonathan Wells writes about evolution is wrong," http://www.ncseweb.org/icons/icon-4haeckel.html.

10 Eugenie C. Scott, "Evolution: Fatally flawed iconoclasm," *Science* (Jun 22, 2001):2257-2258.

11 "[C]ommunity in embryonic structure reveals community of descent," says Darwin in the 1st, 2nd, and 3rd editions of the *Origin*. In the fourth edition (1866), Darwin added the following passage to his discussion of embryology: "[B]ut discommunity in embryonic development does not prove discommunity of descent, for in one of two groups all the developmental stages may have been suppressed, or may have been so greatly modified as no longer to be recognized, through adaptations, during the earlier periods of growth, to new habits of life." Charles Darwin, *The Origin of Species, by Charles Darwin: A Variorum Text*, ed. Morse Peckham, (Philadelphia: University of Pennsylvania Press, 1959):703.

* To explain what might cause these adaptations in embryos, some evolutionary biologists invoke "macromutations," large-scale changes in form that occur in one generation. One such biologist, the late University of California at Berkeley geneticist Richard Goldschmidt, believed that such macromutations could produce what he called "hopeful monsters." We'll tell you more about that when we talk about the modern version of Darwin's mechanism later in this book.

THE JURY TAKES A RECESS

Many modern biologists are embarrassed by the outcry over inaccurate drawings that still appear in some widely-used biology textbooks. As you, as the juror, weigh the case for and against the Darwinian understanding of the history of life, you might be tempted to hand in your verdict now. But that would be premature.

Stephen Jay Gould (who made the critical comment at the end of the Reply section in the last chapter) was well aware of some problems with the embryological evidence, yet remained a staunch advocate of the theory of Universal Common Descent. He wasn't alone. Despite contemporary challenges, a large number of biologists continue to assert the fundamental accuracy of Common Descent. As the late Professor Gould often reminded his readers, the case for Universal Common Descent is not based on a single isolated argument, but upon "a concilience" (or coming together) of many lines of evidence.*

Remember, too, that a good juror should keep an open mind until all the evidence has been examined. We still have one more class of evidence to consider in the case for Common Descent. And we haven't even begun to consider natural selection yet. ■

* Stephen Jay Gould, "Darwinism Defined: the Difference Between Fact and Theory." *Discover* (January 1987):64-70.

BIOGEOGRAPHY

BIOGEOGRAPHY: CASE FOR

Take a look out of the nearest window. See any zebras? How about elephants? If you live in North America, probably not—except at the zoo. Yet, if you lived in a village in the Serengeti, you might see all of them regularly. Animals (and plants) are distributed across the Earth in distinctive geographical patterns. The study of these patterns is called Biogeography.

Darwin spent two chapters in *The Origin of Species* examining what he called "Geographical Distribution" or what today we call *Biogeography*. Darwin thought the patterns of distribution of plants and animals could be best explained by descent with modification from a common ancestor. He used the plants and animals of the Galápagos Archipelago to illustrate his point.

The Galápagos Archipelago, a group of islands about 600 miles off the coast of Ecuador, is home to many species of land birds, marine reptiles, and plants. Darwin noticed that many of the species on the different Galápagos Islands closely resembled each other. For example, Darwin observed three distinct species of mockingbirds on three separate Galápagos Islands. He noticed that these species were far more similar to each other than they were to any other species of mockingbirds found around the world. But why should this be?

Darwin also wondered why organisms found in the Galápagos resembled organisms found in South America. As he put it, "almost every product of the land and water," both plant and animal, "bears the unmistakable stamp of the American continent." If organisms were created to live just where we find them, he wondered, "Why should the species which are supposed to have been created in the Galápagos Archipelago, and nowhere else," resemble organisms found in South America.

Contemporary evolutionary biologists give the same answer to these questions that Darwin did. We see biological *similarities* because the Galápagos animals are descendants of animals that migrated from South America. We see biological *diversity* because the Galápagos animals adapted in different ways after arriving in their new environments. Also, species like the Galápagos mockingbirds are more similar to each other than they are to other mockingbirds because they evolved from the same original mockingbird species after it migrated to the Galápagos.

But how can we be sure that these species didn't arise independently? Let's take a closer look at the Galápagos Islands and the organisms that we find on them.

The Galápagos Islands are young (geo-

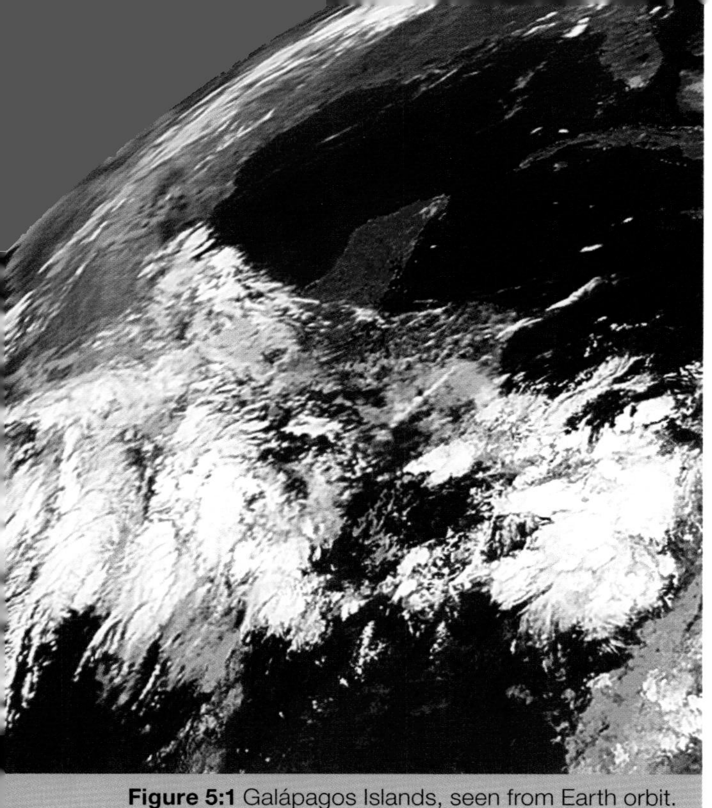

Figure 5:1 Galápagos Islands, seen from Earth orbit.
Photo courtesy of NASA

logically speaking), and are isolated from the mainland. In spite of this, many species on the mainland of South America have representatives in the Galápagos. But the Galápagos lacks many of the mainland species. Why is that significant? It suggests that Galápagos species migrated from the mainland, not the reverse. If you look carefully at the species on the Galápagos, you'll see they have something in common. The Galápagos inhabitants are capable of traveling large distances over water. The Galápagos animals can fly, swim or float; the Galápagos plants have seeds that can be carried by birds, wind, or water.

Large land mammals, on the other hand, are not equipped to cross from the South American mainland to the islands. And sure enough, those are the ones we *don't* find on the Galápagos. Conclusion? The most likely explanation for the distribution of species is migration and subsequent adaptation. The Galápagos animals migrated from South America and evolved to adapt to their new environment.

Contemporary evolutionary biologists point to other biogeographical evidence in support of this view. For example, in the Hawaiian Islands, there are hundreds of species of the fruit fly genus *Drosophila*. These species are not found anywhere else in the world. Why are there so many different species of fruit fly in such a remote place? As the National Academy of Sciences argues, "The biological explanation for the multiplicity of related species in remote localities is that such great diversity is a consequence of their evolution from a few common ancestors that colonized an isolated environment."*

Some biologists also point to the concentration of marsupial mammals in Australia and South America as evidence of descent from a common ancestor. Why are marsupials mainly found on those two continents? Proponents of Common Descent offer this explanation. The first mammals with the marsupial's distinctive mode of reproduction arose on the ancient southern super-continent of Gondwanaland. Later, after this great land mass broke up into separate continents, the ancestors of the modern marsupials were separated from other mammals and evolved in isolation on the new continents of Australia and South America. The distinctive Australian marsupials—kangaroos and koalas, for example—are the result of that evolutionary process in an isolated locale.

* *"Science and Creationism: A View from the National Academy of Sciences,"* Second Edition. Washington, D.C.: National Academy Press (1999):17.

Biogeography: Reply

You may be surprised to learn that most critics of neo-Darwinism agree with many of the arguments you have just read. At least, up to a point. Few biologists today would argue over whether the different species of Galapagos mockingbirds (for example) descended from a common ancestor.

Darwin's discussion of biogeography may have an odd sort of ring to modern ears. This is largely because Darwin was using this evidence to challenge a theory that was popular in his day, but is almost unheard of now: the fixity of species. The *fixity of species* was the idea that each species is fixed in its physical form from which it doesn't change (at least not enough to constitute a new species) and placed in its current habitat from which it doesn't move (at least beyond significant geographic barriers such as mountain ranges or oceans). Nowadays, the idea of the fixity of species isn't even a blip on the radar. Very few scientists, even those who are critical of neo-Darwinism, disagree with Darwin's migration/adaptation scenario.

For example, most modern critics of neo-Darwinism accept the idea that all the mockingbirds of the Galápagos have a common ancestor. In their view, the evidence *does* support the idea that these birds have changed in response to their environment (Evolution #1), but it does *not* show that all creatures everywhere have a single common ancestor (Evolution #2). And that's the rub. These scientists accept that plants and animals of the Galápagos were transported or migrated to the islands and then adapted in some ways to their new environment. They point out, however, that migration and adaptation does not equal macroevolutionary change.

Critics say that the evidence of the minor variations that we see in both Galápagos plants and animals and the Hawaiian fruit flies favors neither the Darwinian view nor its modern competitors. The evidence is just as consistent with a polyphyletic view (the orchard picture of the history of life, in which only minor variation has taken place) as it is with the monophyletic view (the single tree picture of the history of life, in which macroevolutionary change has taken place).

In other words, many critics contend that at first glance, the evidence from biogeography is inconclusive.

Some critics go further, saying that the patterns of geographical distribution we observe

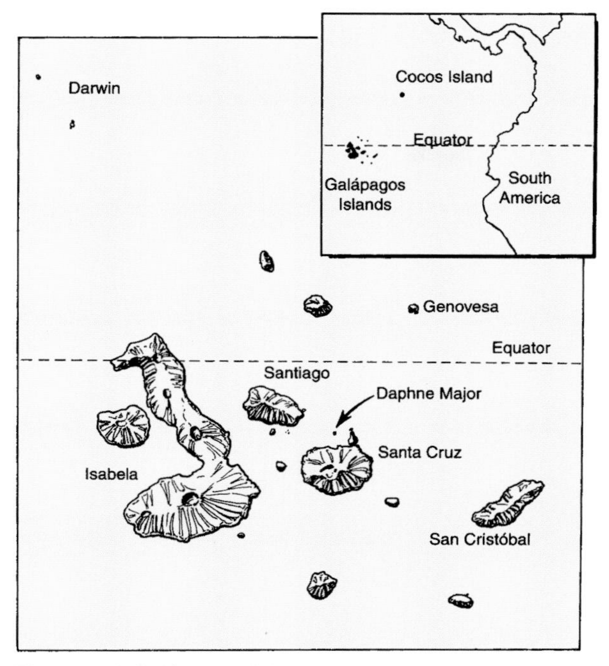

Figure 5:2 Galápagos Islands *Illustration courtesy of Jody Sjogren*

Three Definitions of Evolution

- Evolution #1: Change over time.
- Evolution #2: Universal Common Descent.
- Evolution #3: The Creative Power of Natural Selection.

raise deeper questions about Universal Common Descent. If Universal Common Descent is true, it must have a mechanism that can produce macroevolutionary change—that can transform one type of animal into a fundamentally different type of animal. Yet critics note that the examples of mockingbirds in the Galápagos and fruit flies in the Hawaiian Islands show only small-scale variations in existing traits. Further, some geneticists think that these changes have occurred because the populations of these birds and fruit flies became isolated, and lost genetic information over time. There are many examples of isolated islands that are home to flightless birds and insects that have clearly lost some of the genetic information necessary to produce the traits possessed by their ancestors).* This poses a logical and practical problem, say the critics. Large-scale macroevolutionary change requires the addition of new genetic information, not the loss of genetic information.

Since critics of the argument from biogeography see no evidence of large-scale change, or of a mechanism that can produce the new genes needed to cause such change, they doubt that the biogeographical distribution of animals supports *Universal* Common Descent.

Critics of the marsupial argument insist that it, too, fails to establish Universal Common Descent or even the common descent of all marsupials. At best, it shows that various groups of marsupials first originated in the same general area in the Southern Hemisphere and were then distributed more widely as the Southern conti-

* Isabelle M. Côté & William J. Sutherland, "The effectiveness of removing predators to protect bird populations," *Conservation Biology* 11 (April 1997):395.

Gustav Paulay, "Biodiversity on oceanic islands: Its origin and extinction," *American Zoologist* 34 (1994):134-144.

nents separated from one another. But even this is questionable, some critics say. They point out that marsupials are *not* restricted to the southern continents of Australia and South America. Marsupials such as the opossum live in the northern hemisphere. And, in a recent development, paleontologists have unearthed the oldest marsupial fossil of all... in China.

If the ancestors of the marsupials originated in the Southern Hemisphere, why has the oldest known member of the group been discovered in the Northern Hemisphere? Puzzles like this have led some critics to question how much we really know about where marsupials originated and whether the argument from biogeography provides any evidence for Universal Common Descent. Critics say that the worldwide distribution of marsupials provides evidence for only two things: continental drift and migration.

By the way...

Scientists who held to the fixity of species were not the only ones to challenge Universal Common Descent during the 19th century. Georges Cuvier (a highly respected naturalist and comparative anatomist) knew that changes in the environment could force organisms to move to new habitats, sometimes far from their original range. These migrating organisms would have to adapt to their new environment. Cuvier accepted that migration and adaptation would alter the features of species. Nevertheless, he doubted that species could undergo *unlimited* change, and did not accept that all species shared a common ancestor. Many modern critics of neo-Darwinism share this view.

BIOGEOGRAPHY: FURTHER DEBATE

Frankly, at one level there may not be much further debate about biogeography. Unless somebody, somewhere, makes an astounding discovery on one side or the other, the issue is likely to remain exactly where it is.

Those who accept Universal Common Descent will probably continue to say that the evidence from biogeography supports their view. Since different populations of Galápagos mockingbirds and Hawaiian fruit flies have clearly changed over time and descended from a common ancestor, the distribution of these animals seems to support the possibility that all organisms have descended from a common ancestor. For their part, dissenters will continue to point out that the evidence is completely consistent with other views of the history of life, in which small-scale changes in form and features do occur within separate but disconnected groups of organisms.

Summary

What does the evidence from biogeographical distribution tell us about the theory of Universal Common Descent? Is the history of life best represented as a single branching tree, or as an orchard of separate, disconnected trees? We have seen that scientists sometimes disagree about how to interpret the various classes of evidence we have examined. Scientists disagree about what paleontology, or embryology, or biogeography tells us about the history of life.

As our recent discussion of biogeography shows, some of these disputes arise because scientists disagree about how much change biological mechanisms can actually produce. The single continuous branching tree of life that Darwin envisioned requires a mechanism that can produce a lot of change—a mechanism that can eventually transform one-celled organisms at the base of the tree of life into all the complex plant and animal forms that we see today.

A scientist who thinks that mechanisms of biological change can only produce limited or small-scale change will, therefore, tend to doubt this single-tree (monophyletic) picture of the history of life. Conversely, a scientist who thinks there *is* a mechanism sufficient to produce evolutionary change on a large scale will tend to favor the monophyletic picture of life's history.

How much creative power do evolutionary

mechanisms possess? That is a key question in the current controversy about Darwinian evolution. Since understanding this mechanism is vitally important to our understanding of the history of life, and since Darwin and modern neo-Darwinists have claimed that natural selection can indeed produce large-scale biological change, the next part of this book will examine the arguments for and against the creative power of this mechanism.

THE CREATIVE POWER OF NATURAL SELECTION

Darwin himself described *The Origin of Species* as "one long argument" for his theory of descent with modification. By "descent," he meant Universal Common Descent. As you recall, the original and the contemporary versions of Darwinian theory claim that all organisms descended from a common ancestor. Thus far in the book, we've examined the five main lines of reasoning that Darwin and/or his modern advocates use to support that part of evolutionary theory.

By "modification" Darwin meant the creative power of natural selection acting on random variations, which brings us to the second part of Darwinian theory. In *The Origin of Species*, Charles Darwin proposed a mechanism that he thought could cause living organisms to change, and cause new forms to arise from earlier forms of life. He called his mechanism Natural Selection. In this section of the book, we will present the three main arguments used to support the creative power of natural selection. Darwin thought natural selection was a creative force because it acted on naturally arising variations in the features of organisms. Neo-Darwinists agree, but emphasize a special kind of variation called *mutation*, unknown in Darwin's time. The next two chapters will examine these two important ideas. ■

NATURAL SELECTION

Natural Selection: Case For

As we have already discussed, Darwin's theory attempts to establish that all living organisms have descended from a common ancestor. Of course, characterizing the history of life in this way means that life has changed dramatically along the way. But, what caused those changes?

Trying to identify the "cause behind the change" raises another scientific issue. Is there a natural mechanism that is capable of generating new biological structures and new living forms? Is there a natural process powerful enough to produce the extraordinary change in living organisms, as depicted by the Tree of Life?

Darwin was convinced there was. A second part of his theory described this mechanism, which he called natural selection, and developed an argument to demonstrate its creative power. In the *Origin*, Darwin explained how the large-scale changes—required by his theory of Common Descent—could take place over the course of geologic time by means of natural selection.

Looking for a Mechanism

Although we now tend to attach the word "evolution" to his name, Charles Darwin did not invent the idea. Long before he set sail on the *Beagle* voyage, others had proposed theories about how organisms arose naturally and changed from one type to another. But as far as Darwin was concerned, all previous evolutionary theories suffered from the same overwhelming defect. They had no mechanism. Darwin realized that unless he could identify a mechanism—a process that was capable of producing changes in form—his own theory would remain simply an interesting speculation.

A *mechanism* is just a fancy way of saying a process by which something happens: a "cause." For example, airplanes don't rise by magic. When air moves past a surface of a particular shape, it creates lift. The faster the air moves over and under that surface, the greater the amount of lift. Enough speed creates sufficient lift for a plane to leave the ground. Not magic, a mechanism.

Darwin rejected evolutionary theories that appealed to mysterious events. It wouldn't be enough, he wrote, to say that "after a certain unknown number of generations, some bird had given birth to a woodpecker."[1] Woodpeckers display particular specialized features not shared with other birds. The origin of these features requires careful explanation in terms of an observable, biological process.

Darwin thought that he had found such a mechanism. He realized that to explain the origin of new forms, his mechanism would have to do two things.

First, the mechanism would have to be able to modify structures and features that already existed in organisms. Second, it would have to preserve the small changes that had been made, so that they could add up to big changes over time.

Darwin thought that the mechanism of natural selection could do this, and neo-Darwinian biologists today agree. In fact, many leading evolutionary theorists insist that natural selection is "... the main process by which evolutionary change comes about." [2]

Natural Selection: Three Necessary Conditions

Natural selection is a three-step process, and each step has a logic that is easy to understand. Each step is important. You could think of the process as a three-legged stool, which can stand only when all three legs are present. In the same way, each of the three steps in the process of natural selection provides something necessary to produce significant biological change. Let's look at those three steps.

Step 1: Variation

Darwin noticed that *all organisms vary.* Think about the members of your own family. While you probably resemble your parents (and siblings) in many ways, you also exhibit many distinctive characteristics of your own. No two organisms are exactly alike. Cows from the same herd are not exactly alike, and even puppies in the same litter are not exactly alike. According to Darwin, these variations provide the raw material for the changes in biological form that can occur over time. Neo-Darwinists think random changes in DNA (called mutations) provide this raw material. Variation is the first leg of the stool. But variations alone are not enough.

Step 2: Heritability

Darwin observed that many variations could be passed on from one generation to the next. That is, the variations were *heritable.* Although Darwin didn't know exactly how new variations arose, or how they were passed on from one generation to the next, he did notice that pigeons with fantails (for instance) were likely to have offspring with that same characteristic. (Domestic breeds, in general, provided Darwin with many examples of heritable variation.)

He reasoned that if permanent change in species were to occur, the newly arising variations must be transmitted faithfully, through reproduction, from the parent to the offspring generation. This, then, is the second leg of the stool—the second necessary condition for natural selection to produce biological change: the *variation must be heritable.*

Step 3: Differential Reproduction

Finally, Darwin argued that life was characterized by a "struggle for existence." He noted that in this competition among organisms to survive and reproduce, some variations gave a *competitive advantage* to the organisms possessing those traits. In these cases, the variation directly affects which members of the group survive, and which do not. The "advantaged" organisms would pass this trait on to their offspring. Over the course of several generations, the advantageous features would appear in more and more individuals in the surviving population, causing the population as a whole to look different than it did at the beginning of the process.

Let's say, for example, that a few field mice need less water than the other field mice in the same environment. Let's call these the "hardy" mice. During a prolonged drought, the hardy mice would stand a better chance of surviving.

Nature "selected" the hardy mice. Needing less water gave these mice a "competitive advantage," and led to their "reproductive success." This is an example of the third leg of the stool—differential reproduction—the final necessary condition for natural selection to produce significant change.

If	A	the drought persists,
and	B	the hardy mice are more likely to reproduce (because they're more likely to be alive),
and	C	the hardiness trait is heritable,
then	D	many individuals in the next generation will require less water than the mice that existed before the drought began.

And that's the three-part logic of natural selection. To summarize, natural selection requires (1) variation, (2) heritability, and (3) reproductive difference.* Darwin and his modern followers say that if all three of these elements are present, natural selection can produce significant biological change.

But can this three-step process construct organs as complex as an eye, or structures as complex as a wing? Darwin thought that it could, if given enough time. He made two arguments to build his case for the adequacy of natural selection as a creative mechanism. (Modern evolutionary biologists use one additional argument.) One argument asserts that the small-scale changes we see in species today can be extended (over long periods of time) to explain large-scale changes. We're going to call this argument "microevolution, extrapolated." We'll explain this argument in a moment. But first we're going to look at Darwin's other argument, which compares natural selection to artificial selection.

Analogy to Artificial Selection

Darwin's first argument for the creative power of natural selection came from the familiar practice of animal breeding. He pointed out that animal breeders routinely change the characteristics of their own flocks and herds.

Suppose a shepherd has a flock of sheep. Some, he notices, have especially thick wool. He removes the woolliest rams from one pen and places them together in another pen. He allows them—and no others—to mate with the ewes of his flock. After the lambs are born, he once again separates out only the woolliest from among

* Other biologists use terms like "differential survivability" and "differential reproduction."

them. And so on. Predictably, after a number of generations, the characteristics of his whole flock will have changed. How? They will all be woollier.

By repeating this process (called selective breeding) over several generations, one could modify the characteristics of the whole breed. "The key is man's power of accumulative selection," wrote Darwin. "Nature gives successive variations; man adds them up in certain directions useful to him." [3]

Is it possible that something like this process occurs in nature—only without any intelligence to guide it? Darwin thought so.

Think about our sheep example again. This time, instead of a shepherd selecting woolly sheep according to his own criteria, suppose there was an environmental change that prevented all but the woolliest from breeding. For example, what if there were a series of unseasonably cold winters? Darwin reasoned that natural processes like this could change a population in the same way. Only the woolliest sheep would survive a very severe winter, so only they would leave offspring. The woolliest sheep—the ones with the woolly advantage—would pass that same variation to the next generation. Given enough time, Darwin thought, small-scale changes like these would add up generation after generation, eventually resulting in a very large change in the population. Darwin was confident that what breeders can do in a short time, nature could do over a long time. Darwin was convinced that the power of natural selection was effectively unlimited. This brings us to Darwin's second argument.

Microevolution, Extrapolated

Darwin knew that both breeders and nature itself could produce small changes within a population in a short time. Biologists now call this small-change-in-a-short-time "microevolution."

Because of the length of time that organisms have existed on earth, Darwin thought that naturally-occurring small changes could add up to big changes over time. This was his second argument, the argument from microevolution, extrapolated.

Extending small to big is a handy way to define "extrapolate." For example, suppose you planted a pine seedling and decided to chart its growth over the years. After one year, the seedling grows to twelve inches in height; after three years, it is nearly a yard high. By extrapolating (i.e., projecting or extending) that rate of growth, you might predict that after five years the tree will be about five feet tall.

Figure 6:1 Pine seedling growth, extrapolation

Darwin's extrapolation followed the same logic. He noticed that many species displayed small-scale, naturally occurring changes. He then projected how species might be modified if these changes accumulated over time. Darwin's idea was that if this process of collecting, adding up, and extending changes went on indefinitely, it would ultimately produce large-scale change—the new branches and the branching action described by his Tree of Life.

Biology textbooks cite two classic examples to support the claim that natural selection can produce small-scale change over a short time. In Darwin's theory, such small scale changes should accumulate to produce large-scale change over a long time. Let's consider these two examples.

Finch beaks

In 1977, two biologists, Peter and Rosemary Grant, witnessed the kind of small-change-in-a-short-time that many consider the starting place for Darwinian evolutionary change. The Grants studied one species of finch on one of the Galapagos Islands. After a severe drought in 1977, the Grants observed that 85% of those finches had died, and that the ones that survived were mainly those with the largest beaks. Apparently, birds with larger beaks were the only ones that could crack the few remaining extra-hard seeds that had made it through the drought. The average beak size was now about 5% larger, a definite case of a small change in a short time.[4] The Grants estimated that about twenty such droughts could increase the average beak size enough to produce a new species of finch. They argued that the microevolutionary changes they observed in a short time demonstrated natural selection's capacity to produce much greater change over long periods of time, as required by Darwin's theory.[5]

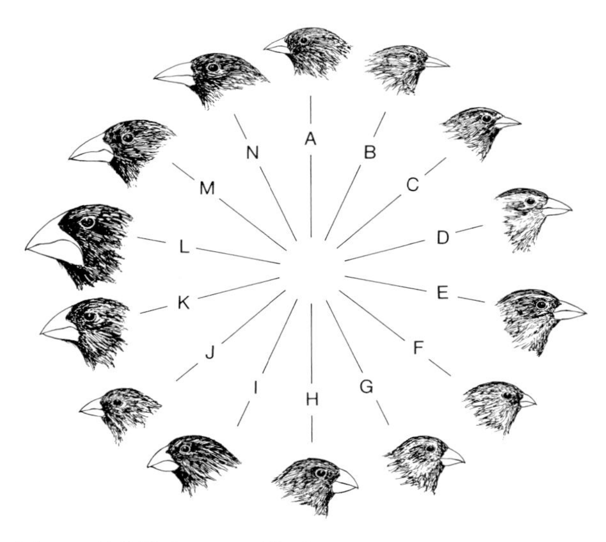

Figure 6:2 Galapagos Finches *Illustration courtesy of Jody Sjogren*

Peppered Moths

Evolutionary biologists point to another classic example of natural selection producing micro-evolution—the changes in the coloration of peppered moths.

Here's what the textbooks say happened. During the 1800s, the population of peppered moths in England shifted from being made up of mostly light-colored moths to a population of mostly dark-colored moths. In pre-industrial

Figure 6:3 Peppered Moths *Used by permission from Wolfson College, Oxford University*

England, many trees were covered with lichens (organisms that grow on rocks and trees). The lighter moths were more numerous during this time. They were well camouflaged, since they blended in with the light-colored mottled appearance of the lichens. The dark moths were easier for hungry birds to see, so the dark moths were dinner, while the light moths were relatively safe.

Then came the Industrial Revolution, when coal-fired factories darkened the tree trunks with soot. The population of dark moths increased during this period because they were now better camouflaged on the pollution-darkened tree trunks. The light colored moths, on the other hand, were more visible against the dark background. Now the dark moths were safe, while the light moths were lunch. Consequently, the dark moths became more numerous than the light moths.

In the 1950s, to prove this hypothesis of "selective predation," British physician Bernard Kettlewell released both light and dark peppered moths onto tree trunks in both polluted and unpolluted woodlands.[6] He watched as birds readily ate the more visible moths. Kettlewell called this example of natural selection "Darwin's missing evidence." Modern neo-Darwinists agree, because Kettlewell's experiment shows how natural forces—a combination of a change in the environment and the action of natural predators—changed the characteristics of the moth population over a short period of time. These are the types of changes required—and predicted—by Darwin's theory.

Evolutionary biologists argue that the examples of finch beaks and peppered moths show that natural selection has the power to produce change. In Darwinian theory, if changes such as these accumulate over time, they can produce the tree-like pattern that Darwin thought characterized the whole history of life. Evolutionary biologists summarize their argument this way: natural selection produces small change in a short time. As small changes add up, new species arise. One species gives rise to two. These continue to diverge, and evolve into new genera, and so on.* This, they conclude, is the way the Tree of Life grows. Tiny twigs (small variations) eventually become great branches (the major differences between organisms). Moreover, given enough time, it seems reasonable that natural selection could produce all the diversity of life we see today.

Ultimately, say evolutionary biologists, the accumulation of changes, continuing over time, can produce fundamentally new forms of life. As Darwin himself said, "I can see no limit to the amount of change... which may have been effected in the long course of time through nature's power of selection."[7]

* That's the plural of genus: one genus, two genera.

NATURAL SELECTION: REPLY

Most critics of Darwin's argument would agree that nature can "select" for successful adaptations or variations. Most would also agree that natural selection can produce small-scale changes (Evolution #1). Nevertheless, critics contend that natural selection's power to change a species is limited; it does not have the almost boundless power the theory requires.

Critics of Darwinian theory readily agree that both nature and human breeders can produce some change within populations. Artificial selection, intentionally mating animals that have certain traits with the goal of enhancing those traits, is a well-established process.

Artificial Selection Analogy: Reply

For the critic, the question is not whether sheep can become woollier sheep; the question is whether sheep can eventually become sheepdogs... or horses... or camels. In other words, can natural selection transform one form of life into a fundamentally different form of life? Can natural selection really produce the large-scale changes that Darwin's theory requires? Critics say it cannot, and point to examples like the following to make their case.

Hitting the Limits

Try to imagine a dog tiny enough to fit into an eyeglasses case, or large enough to rival a horse. Of course, this is comical. But the reason it's humorous is that we know that this sort of thing happens only in cartoons. But why is that? Could it be that there are limits to how much an animal can change? Critics think so.

Figure 6:4 shows an array of dog breeds with strikingly different sizes and shapes. But as different as these breeds are, the differences still fall within limits. No one has ever bred a dog lighter than a few pounds, or heavier than about 150 pounds, despite thousands of years of selective breeding. Critics say that the experimental evidence reveals definite, discoverable limits on what artificial selection can do.

They point out that animal breeders hit limits all the time. Breeders have tried for decades to produce a chicken that will lay more than one egg per day. They have failed.[8] Horse breeders have not significantly increased the running speed of thoroughbreds, despite more than 70 years of trying.[9] Darwin's theory requires that species have an immense capacity to change, but the evidence from breeding experiments shows that there are definite limits to how much a

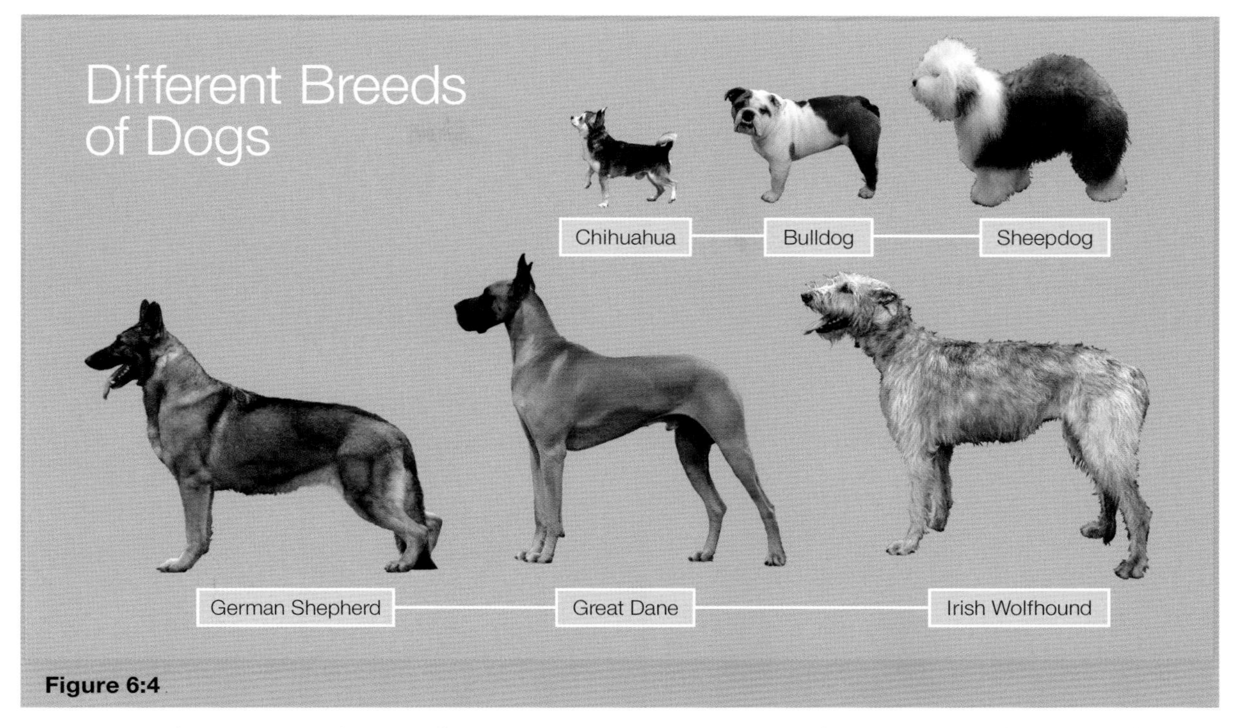

Figure 6:4

species can change, even when intelligent agents (the breeders) are doing the selection intentionally, trying to maximize certain traits.

Critics point out another problem. Intense programs of breeding (and inbreeding) frequently increase the organism's susceptibility to disease, and often concentrate defective traits. Breeders working with English bulldogs have strived to produce dogs with large heads. They have succeeded. These bulldogs now have such enormous heads that puppies sometimes have to be delivered by Cesarean section. Newfoundlands and Great Danes are both bred for large size. They now have bodies too large for their hearts and can suddenly drop dead from cardiac arrest. Many Great Danes develop bone cancer, as well. Breeders have tried to maximize the sloping appearance of a German Shepherd's hind legs. As a result, many German Shepherds develop hip dysplasia, a crippling condition that makes it hard for them to walk.[10] When breeders try to force a species beyond its limits, they often create more defects than desirable traits. These defects impose limits on the amount of change that breeders can ultimately produce.

Darwin's theory states that the unguided force of natural selection is supposed to be able to do what the intelligent breeder can do. But even a process of careful, intentional selection encounters limits that neither time nor the efforts of human breeders can overcome. Consequently, critics argue that by the logic of Darwin's own analogy, the power of natural selection is also limited.

Darwin' theory requires that species exhibit a tremendous elasticity—or capacity to change. Critics point out that this is not what the evidence from breeding experiments shows.

Microevolution Extrapolated: Reply

As we have seen, breeders aren't the only entities that can produce small change in a short time. Nature causes changes, too. Most critics of Darwin's theory agree that natural selection can

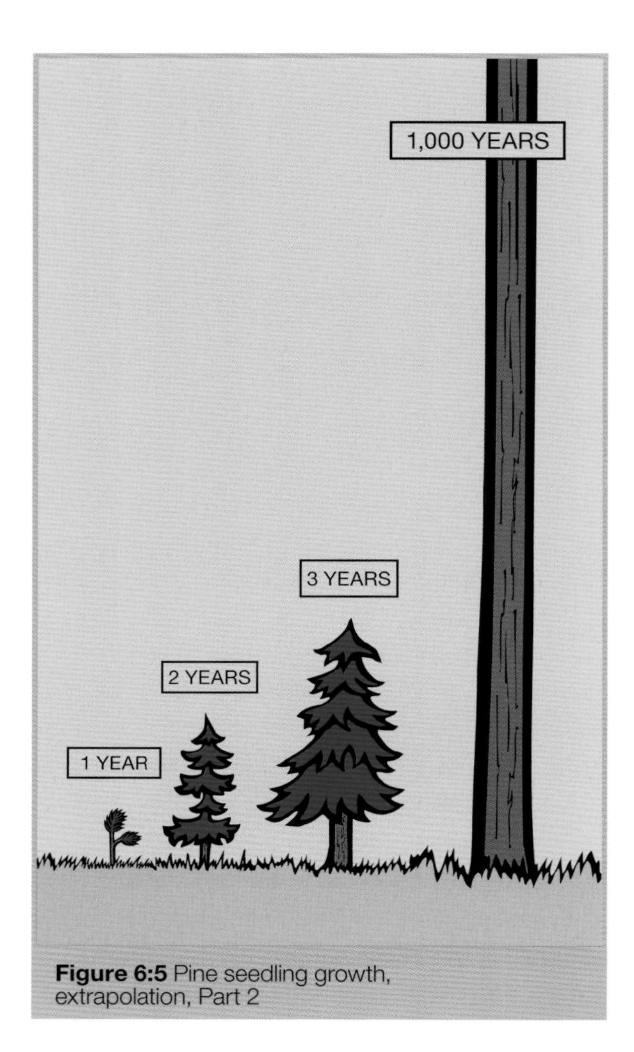

Figure 6:5 Pine seedling growth, extrapolation, Part 2

produce small-scale changes. Nevertheless, they argue that, just as the breeder's power to change a species is limited, the power of natural selection is limited, as well.

Remember our pine tree growth chart from a few pages ago? After one year, the seedling had grown to twelve inches in height, and after three years, it had grown to nearly a yard high. We extrapolated that rate of growth, and predicted that after five years the tree would be about five feet tall. So far, so good. The problem arises when we extrapolate indefinitely into the future. One might assume that after 1,000 years, the seedling will have grown into a tree 1,000 feet high. But it won't grow that high, or anything close to it, even if it lives that long. Extrapolating the growth rate of a seedling indefinitely does not give a biologically reliable answer because there are real limits on the process of growth.

Similarly, critics contend that there are real limits to microevolutionary processes.[11] Far from proving an ability to produce extraordinary change, each of the previously mentioned textbook examples actually illustrates the opposite: that natural selection's capacity to produce change is limited.

Finch Beaks

First, let's review the example of the finch beaks and the Galapagos drought, to see what other scientists have to say about this classic Darwinian story. Critics of this argument acknowledge that the average depth of finch beaks (top to bottom) did increase after the drought of 1977. Yet, they question whether this illustrates the creative power of microevolutionary processes over time. For openers, they point out that nothing new was created. Big beaks and small beaks were present in the population before the drought, during the drought, and after the drought. It would be true to say that the bigger-beaked birds within the population were more likely to survive the drought. But that is not the same as saying that

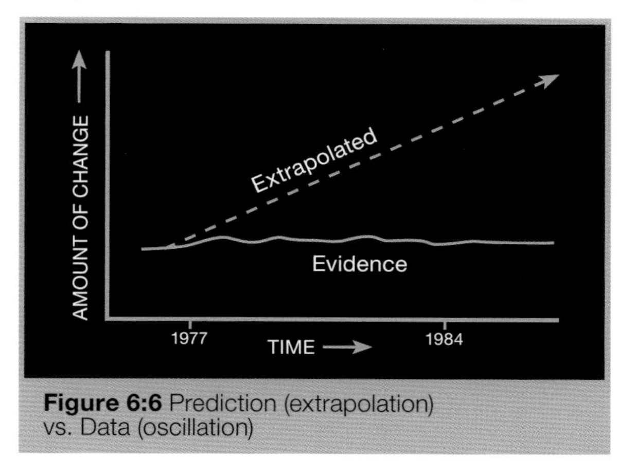

Figure 6:6 Prediction (extrapolation) vs. Data (oscillation)

the population acquired bigger beaks. No new traits arose. The only thing that changed was the *proportion* of big beaks to small beaks.

But there is another problem with using the Galapagos finches to illustrate the capacity to produce large-scale change. After the heavy rains of 1983, the depth of the average finch beak went back to its pre-drought size, and the so-called "evolutionary change" was reversed.

Even though beak length in another species showed a slight net increase, the facts suggest that microevolutionary processes produce only minor variations within definite limits. They do not show that such processes can produce fundamentally new organisms or structures. Nowhere in the finch beak story does a new family, genus, or even species emerge.

In fact, after the rains returned, the Grants noticed that several separate species of finches were interbreeding.[12] Not only were no new species springing forth, but existing varieties actually seemed to be merging. Critics therefore conclude that the finch beak example of microevolution actually suggests that biological change has limits.[13]

Peppered Moths

But what about the case of the peppered moths, with camouflaged moths hiding on tree trunks? In this case, too, skeptics say it's time to take a closer look. They contend that the story contains errors of fact and errors in how the facts are interpreted.

They first point to the problems of interpretation. Critics question whether the peppered moth story shows that microevolution can eventually produce large-scale change. They point out that nothing *new* emerged. Both light moths and dark moths were present in the pre-industrial countryside, in the industrial period, and in the more environmentally-conscious present. They point out that only the *proportion* of light- and dark-colored moths in the population shifted back and forth over time. No new trait appeared in the existing peppered moths, no new variety of peppered moth appeared, and no new species of moth arose. Once again, critics say that what the story actually shows is fluctuation within pre-existing limits. The story does not show that natural selection created anything fundamentally new... not moths, not trees, not birds. Thus, critics say the peppered moth story does not show the unlimited creative power of natural selection.

But there's more. Many research scientists contend that the experiment itself is invalid. Remember that Kettlewell's experiments were supposed to show that natural predators preyed selectively on those light-colored moths that landed on pollution-darkened tree trunks. This process of "selective predation" supposedly caused the shift in the populations of light-and dark-colored moths: a clear case of natural selection in action.

But there are at least two problems with this experiment. First, biologists now know that peppered moths are night-fliers. But releasing the moths at night would have made observation difficult. So, to better observe the birds eating the moths, the experimenters released the moths during the daytime, when the moths were normally asleep. But releasing the moths when they were sleepy and sluggish tells us nothing about what would happen in the wild.

The second problem is where the moths were placed. Peppered moths in the wild normally find resting places high up in the canopy of the trees—not on the tree trunks where the experimenters placed them.* The upshot is that the experimenters released peppered moths that were sleepy and disoriented, placing them on tree

* So, what about all those amazing pictures of camouflaged moths on tree trunks? Most of those moths were manually placed on the tree trunks by the researchers themselves. Some are actually pictures of dead moths that have been pinned or glued to the trunk!

trunks by hand, where they became unnaturally easy targets for predatory birds. Clearly, this does not simulate what takes place in the wild.[14]

So, say the critics, not only does the experiment not show what the story says it's supposed to, the experiment itself is highly questionable.

Why Limitation is Inevitable: The Information Problem

Some critics contend that there's a deeper reason the classic examples of artificial selection and microevolution fail to demonstrate the creative power of natural selection. These processes do not produce the novel biological information needed to build fundamentally new forms of life. Before we go any further, maybe we should explain what scientists mean by the term, "biological information."

You're probably familiar with the general idea of "information," although maybe you've seen it in a different context. Here's an everyday example.

Suppose that your new computer came with only an operating system, a web browser, and a basic text editor installed. However, you want your history report to include pictures you've downloaded from the web. Your computer needs new functionality. What do you do? You buy new software and install it, which provides your computer with an influx of new information (in the form of new lines of software code). New functionality requires more information.

What does this have to do with biology? It turns out that the same principle applies to living things. If a new and different life form is going to arise from an existing one, it doesn't just require more cells, it requires more *kinds* of cells—cells that perform specialized functions. Furthermore, each new cell type requires many new and specialized proteins. But to build new proteins for these new features, you need new genetic information.*

A large portion of the information needed to construct an organism (with its various traits) is stored in the molecule DNA. Some scientists refer to this information as "assembly instructions" or "a genetic program." Just like a computer program, DNA contains the biological equivalent of lines of computer code. Evolutionary zoologist Richard Dawkins states, "The machine code of the genes is uncannily computer-like."[15]

So, biological information is stored in DNA. But where does *new* biological information come from? Critics of neo-Darwinism contend that contemporary evolutionary theory doesn't have an adequate answer for this question. They say that the examples of artificial selection and microevolution in particular do not demonstrate the ability to add new biological information into a population.

To see why they say this, imagine a small group of people. Think about the people in your own family. Chances are your hair and eye color and other physical traits pretty closely match those of your siblings and other family members. That's because the gene pool** of your family is relatively small. Now think about everyone in your neighborhood. This larger group will probably exhibit many features not found in your own family. There will probably be greater variety of natural hair color and eye color. Perhaps everyone in your family is between 5'6" and 6'2" tall, yet in your larger neighborhood there is a 6'8" basketball player and a 4'8" gymnast. Now, if you considered the features of everyone in your ZIP code, you'd probably find an even greater variety. You can think of these larger communities as concentric circles. The larger the circle, the more genetic variability it contains.

* This is an exciting field for research. Scientists are only beginning to understand the role of information in biology.

** The term "gene pool" is an informal way to refer to the total amount of genetic information available to a particular population. The study of changes in the proportion or "frequency" of genes in populations is called "population genetics."

In selective breeding, by contrast, breeders don't want variability. Instead, they intentionally restrict the size of the breeding population to a very small circle, the size of the circle that represents your own family. This enables breeders to identify and enhance a single desirable genetic trait (like thick wool or a large head). This enhancement comes at a cost, however. By restricting the genetic variability, the resulting population has lost genetic information for building certain other traits—traits that may be needed for survival some day. The process of selective breeding limits the extent to which such populations can vary and change. Artificial selection occurs as biological information and genetic traits are lost to the population as a whole. That is why artificial breeding always encounters limits, say the critics.

The same problem affects microevolutionary processes in nature. It would not be unheard-of, for example, for two populations of about 100 finches each to become isolated from a larger flock of 10,000 finches—not as the result of a breeder's intent, but naturally, as the result of undirected migration. The two so-called "daughter" populations will begin to differ. Some traits may begin to show up more frequently in the offspring of one or both of the daughter populations. Why? Because the information for building those traits is now carried by a larger percentage of the members of each population. The converse is also true. Some traits would appear less frequently or not at all, because the information for building those traits may have been lost in one or both of the daughter populations.

Many think this demonstrates the creative power of evolutionary processes, but it doesn't. It's true that certain traits are now being expressed more frequently in one or the other of the isolated populations. It's also true that, as a result, the two new sub-populations will look different from each other (and from the parent population). They will have "evolved" in that limited sense. But these traits—whether dark wings in moths or longer beaks in finches—are not new. The capacity to produce these traits was present all along in the gene pool of the original (large) population.

On the other hand, each daughter population will have lost genetic information necessary for building certain other traits. The total biological information in the gene pool will have decreased, which limits how much the daughter population can vary and change in the future. Ultimately, this means that the isolated "daughter" populations are more vulnerable to environmental stresses (natural disasters or other changes in the environment). For this reason, small isolated populations are great candidates for extinction.

In summary, whether you're talking about artificial selection or about microevolution that occurs naturally, changes in the sub-population take place as genetic information is lost to that population. Here's the rub: producing new organs or new body plans requires new lines of genetic code—*more* information, not less. Not surprisingly, many scientists argue that small-scale microevolutionary change cannot be extrapolated to explain large-scale macroevolutionary innovation.[16] Some would argue that it's illogical to claim that a process that loses information can explain the origin of a new type of animal—a process that needs an influx of new information.

These critics would say that natural selection works well as an editor, but not an author. It has a demonstrated capacity to weed out the failures from among what already exists, but it has not been shown to generate new biological information or structures.

ENDNOTES

1 Charles Darwin, *Origin of Species*, Introduction (3rd ed.)

2 For example, leading neo-Darwinists Dobzhansky, Ayala, Stebbins, and Valentine have said, "According to the theory of evolution… natural selection is the process responsible for the adaptations of organisms, and also the main process by which evolutionary change comes about." *Evolution* (San Francisco: W.H. Freeman, 1977):504.

3 Charles Darwin, *On the Origin of Species* (Cambridge, Mass: Harvard University Press, 1964 [Facsimile of the First Edition, 1859]), Chapter 1, "Variation Under Domestication":30.

4 Directional changes of this sort have been observed in many different species, both animal and plant. Plants that grow near the entrances to metallic ore mines, for instance, can tolerate levels of metal in their environment that would prove toxic to other plants. In these cases, environmental changes brought about by human activity have led to adaptive responses within the species. Adaptive changes in species also occur in response to natural forces, such as drought, the appearance of a new predator, or isolation in a novel environment (such as a cave).

5 Peter R. Grant, *Ecology and Evolution of Darwin's Finches* (Princeton: Princeton University Press, 1986).

see also Peter R. Grant and B. Rosemary Grant, "Unpredictable evolution in a 30-year study of Darwin's finches," *Science* 296 (2002):707-711.

6 H.B.D. Kettlewell, "Selection experiments on industrial melanism in the Lepidoptera," *Heredity* 9 (1955):323-342

7 Charles Darwin, *On the Origin of Species* (Cambridge, Mass: Harvard University Press, 1964 [Facsimile of the First Edition, 1859]):109.

8 I. Michael Lerner, *Genetic Homeostasis* (Edinburgh: Oliver and Boyd, 1954):85, 96-97, 100, 108, and 117.

9 B. Gaffney and E.P. Cunningham, "Estimation of genetic trend in racing performance of thoroughbred horses," *Nature* 332 (April, 21, 1988):722-724.

10 K. Mäki, A.F. Groen, A.E.Liinamo, M. Ojala, "Population structure, inbreeding trend and their association with hip and elbow dysplasia in dogs," *Animal Science* 73 (2001):217-228.

Ross D. Clark and Joan R. Stainer, eds., *Medical and Genetic Aspects of Purebred Dogs* (Edwardsville: Veterinary Medicine Publishing Co.,1983).

A, Heshammer, S.E. Olsson, et al., "Study of heritability in 401 litters of German Shepherd dogs"; *Journal AM Vet Med Assoc* 174 (1979):1012-1016.

Hutt, F.B., "Genetic selection to reduce the incidence of hip dysplasia in dogs"; *Journal AM Vet Med Assoc* 151 (1967): 1041-1048.

Paula F. Moon-Massat DVM, PhD Cornell University, "Risk Factor Analysis of Neonatal Mortality After Cesarean Section in the Dog," Paper presented at the AKC Canine Health Foundation, National Parent Club Canine Health Conference, 1999.

R.G. Mateescu, K.L. Tsai, Z. Zhang, N.I. Burton-Wurster, G. Lust, N.L. Dykes, G.M. Acland, R.L. Qua as, K.E. Murphy, R. Todhunter, "QTL Mapping Using Cross Breed Pedigrees: Strategies for Canine Hip Dysplasia," in *The Dog and Its Genome* E.A. Ostrander, U. Giger, K. Lindblad-Toh, eds. (Woodbury, New York: Cold Spring Harbor Laboratory Press, 2005).

11 Robert L. Carroll, "Towards a new evolutionary synthesis," *Trends in Ecology and Evolution* 15 (January 2000):27-32. [The] explosive evolution of phyla with diverse body plans," Carroll notes, "is certainly not explicable by extrapolation from the processes and rates of evolution observed in modern species, but requires a succession of unique events." Other biologists who have asked whether microevolution can be extrapolated to explain macroevolution including:

Douglas H. Erwin, "Macroevolution is more than repeated rounds of microevolution, *Evolution and Development* 2 (2000):78-84.

David Penny and Matthew J. Phillips, "The rise of birds and mammals: are microevolutionary processes sufficient for macroevolution?" *Trends in Ecology and Evolution* 19 (October 2004):516-522.

Douglas H. Erwin, "The origin of bodyplans," *American Zoologist* 39 (1999):617-629.

David Jablonski, "Micro- and macroevolution: scale and hierarchy in evolutionary biology and paleobiology," *Paleobiology* 26 (2000):15-52.

Daniel W. McShea, "Arguments, tests, and the Burgess Shale—a commentary on the debate," *Paleobiology* 19 (1993):399-402.

George L. Gabor Miklos and Bernard John, "From genome to phenotype," in *Rates of Evolution*, K.S.W. Campbell and M.F. Day, eds. (London: Allen and Unwin, 1987).

George L. Gabor Miklos, "Emergence of organizational complexities during metazoan evolution: perspectives from molecular biology, palaeontology and neo-Darwinism," *Memoirs of the Association of Australasian Palaeontologists* 15 (1993):7-41.

12 Peter R. Grant and B. Rosemary Grant, "Unpredictable evolution in a 30-year study of Darwin's finches," *Science* 296 (2002):707-711.

13 Jonathan Wells, *Icons of Evolution* (Regnery, Washington, DC, 2000):159-175.

14 Judith Hooper, *Of Moths and Men: An Evolutionary Tale* (New York: W.W. Norton & Company, Inc., 2002).

15 Richard Dawkins, *River Out of Eden* (New York: Basic Books, 1995):17.

16 George L. Gabor Miklos, "Emergence of organizational complexities during metazoan evolution: perspectives from molecular biology, palaeontology and neo-Darwinism," *Memoirs of the Association of Australasian Palaeontologists* 15 (1993):7-41.

Stephen C. Meyer, "The origin of biological information and the higher taxonomic categories," *Proceedings of the Biological Society of Washington* 117 (2004):213-239.

Scott F. Gilbert, John M. Opitz, and Rudolf A. Raff, "Resynthesizing evolutionary and developmental biology," *Developmental Biology* 173 (1996):357-372.

K.S. Thomson, "Macroevolution: The morphological problem," *American Zoologist* 32 (1992):106-112.

NATURAL SELECTION AND MUTATION >

NATURAL SELECTION AND MUTATION: CASE FOR

Remember that when we first introduced Natural Selection, we described it as a three-legged stool. (The legs are variation, heritability, and reproductive advantage.) Each of the three "legs" provides something required by the process of natural selection, and contributes to its creative power.

Some scientists see variation as the most important element in the evolutionary process, because it provides the raw material upon which natural selection acts. As evolutionary biologist Søren Løvtrup has noted, "without variation, no selection: without selection, no evolution."[1]

Random Mutation as a Source of Genetic Novelty

As we have seen, there are scientists who doubt that natural selection can produce major evolutionary change. Specifically, they question whether there is a source of new information that can produce new genetic traits—the variations needed to produce lasting biological change.

Defenders of the neo-Darwinian position dispute this critique by offering another argument for the creative power of natural selection. They say that the critics have underestimated the power of another type of variation, unknown in Darwin's time, called mutation. They say that even if the existing gene pool doesn't supply enough information to build a fundamentally new organism, new mutations in the genes can.

Genetic Mutations

A mutation, as you might already know, is a change in the DNA structure. To be more specific, it's a change in the sequential arrangements of the information-bearing bases—the "letters" in the genetic text—of the DNA molecule. Mutations can occur when genes are exposed to heat, chemicals, or radiation.

The most common mutation, as a result of such exposure, happens when a single chemical "letter" in the DNA sequence is changed. Such mutations are called "point mutations." Changing one DNA base pair can cause one amino acid to be substituted for another in the protein being built.

In a "duplication mutation," paired chromosomes swap genetic material as sex cells are being formed. If the two chromosomes are not properly aligned during the swap, one section of DNA is copied twice. One chromosome ends up with two copies of the duplicated genetic

Point Mutation

Let's look at a point mutation in the oxygen-carrying protein called hemoglobin. A point mutation at a single position on the gene causes a valine amino acid to be built where a glutamine should have gone. This substitution distorts the shape of the hemoglobin molecule, causing the red blood cell to become warped. Warped red blood cells do not carry oxygen very well. People who have this condition cannot get enough oxygen into their tissues. The disease called sickle-cell anemia is caused by a single point mutation. The average life expectancy of people who have this disease is about 45 years.*

However, scientists point out that this same mutation can be beneficial to some people, giving them a measure of protection against malaria. Here's how.

The mutated hemoglobin is on the outside of the red blood cell's membrane. As the hemoglobin distorts the red blood cell, giving it the signature sickle shape, it may actually rupture the cell membrane.

You may be asking yourself why on earth this would be beneficial... and in most cases, it wouldn't be. But the malarial parasite spends part of its life cycle within the red blood cell, relying on its high concentration of potassium ions. When the mutated hemoglobin ruptures the blood cell, the potassium is released into the blood stream. With its potassium supply gone, the malarial parasite dies. In this way, a mutation that usually leads to an early death is actually beneficial, giving protection against malaria. ■

*O.S. Platt, D.J. Brambilla, W.F. Rosse, P.F. Milner, O. Castro, M.H. Steinberg, and P.P. Klug, "Mortality in sickle cell disease: Life expectancy and risk factors for early death," *New England Journal of Medicine* 330 (June 9, 1994):1639-1644.

material, while the other chromosome gets none of it.

An "inversion mutation" happens when a section of the genetic text is flipped over so that it reads backwards. Imagine taking 100 letters from one section of this book, flipping it over so it reads backwards, and then inserting it in another section of this book. Major mutational changes such as these do occur in DNA molecules.

Neo-Darwinian biologists contend that mutations can make up for the limited supply of genetic information in the gene pool of a population. Mutations, they argue, provide a new source of variation. In this way, mutations can ultimately produce new traits, anatomical structures, and organisms. Here are two examples that show what mutations can do.

Antibiotic Resistance

In 1928, British physician Alexander Fleming discovered that a species of mold produces a substance—penicillin—that kills many types of bacteria. By the 1940s, penicillin was being used to treat infections in humans. Since then, many other antibiotics have been discovered, including streptomycin, erythromycin, and tetracycline. Most of us have taken one or more of these antibiotics, which help our bodies fight bacterial infection by either killing the bacteria or hindering their reproduction.

However, sometimes the bacteria survive their encounter with the antibiotic, and continue to grow and reproduce. This phenomenon is called "antibiotic resistance."

> ### *The neo-Darwinian Mutation Scenario*
>
> - A genetic program encoded in DNA directs embryonic development.
> - Parents transmit this program to their offspring through reproduction.
> - However, mutations in the DNA sometimes modify this program.
> - Thus, descendants may possess modified structures that are similar—but not identical—to those of their parents.
>
> <<

Two Resistance Mechanisms

How does bacterial antibiotic resistance work? There are two important ways. The first is an enzyme defense system. For example, penicillin kills a bacterium by "gumming up" a molecular machine responsible for synthesizing the bacterium's cell wall. However, bacteria that are resistant to penicillin have an enzyme called "penicillinase" that chemically cuts the antibiotic and renders it harmless. If there is penicillin in the bloodstream, bacteria that don't produce penicillinase die.[2] Bacterial cells that can produce penicillinase survive and reproduce, passing their DNA to their descendants. As a result, the next generation of bacteria is also resistant to the antibiotic.

The second way that bacteria become resistant to some antibiotics is through mutation. Bacteria multiply very quickly. During reproduction, some of their offspring have genes that have been altered by a "copying error" in the DNA.[3] In some cases, this copying error replaces one amino acid in a bacterial protein with a different amino acid. Usually a mistake of this kind would be harmful to a bacterium, but occasionally this actually helps it survive. To understand why, we need to know how this kind of antibiotic works.

When Antibiotics Attack

Antibiotic agents attack critical functions of bacteria, such as genetic information processing. (This includes processes such as DNA replication, RNA synthesis, and protein synthesis). These functions are performed by specific molecular machines made of many protein components.

Antibiotics poison bacteria by homing in on a protein component of one of these essential molecular machines. You've heard of "throwing a monkey wrench into the machine?" That's exactly what's happening here. The antibiotic fits, hand-in-glove, into a vulnerable spot in the structure of a key protein. When an antibiotic latches on to this "target site" (called an *active site*), it prevents the machine component (the protein) from working properly. The bacterium either fails to grow, or dies outright.

However, sometimes a DNA mutation changes the shape of the active site on the target protein. If the mutation changes the shape of the protein enough to prevent the antibiotic from binding to it—but not so much that it destroys the protein's function—then the antibiotic no longer fits, and can't latch on. The result: a single base-pair mutation has given the bacterium a competitive advantage. After the non-resistant cells (those lacking the altered protein) die, the mutant bacterium reproduces to form a large population of antibiotic-resistant bacteria. In a few generations, an antibiotic-resistant strain arises. And mutations are the key.

Neo-Darwinists argue that the development of antibiotic resistance in bacteria is a powerful example of random mutations providing a source of new variation. Mutations produce new genetic

For mutations to provide the raw materials on which natural selection can operate, two things need to happen. First, the "mutants" must be viable (that is, able to survive, and capable of reproducing). Second, the mutation must be heritable. <<

information upon which natural selection can act. The result is a new, fitter strain of bacteria.

Antibiotic resistance has become such a serious health problem (for us, that is) that medical societies now advise doctors not to prescribe antibiotics for minor infections or minor illnesses. Why? If too many types of bacteria develop a resistance to antibiotics, some day there might be a bacteria-borne illness that becomes widespread before we can develop an antibiotic strong enough to kill it.

Structural Mutations: The Four-winged Fruit Fly

Bacteria are single-celled organisms, and the mutation we've just looked at affects only single cells. For major change to occur in more complex, multi-cellular animals, mutations must ultimately affect an organism's shape or structure. Are there examples of such structural mutations? Evolutionary biologists say there are, and point to another striking example of the novel variation that mutation can produce: four-winged fruit flies.

Ordinarily, of course, fruit flies have only two wings, one on each side of their bodies. The four-winged fruit fly is not a variety that normally occurs in the wild. And that's just the point. By inducing mutations through genetic manipulation, scientists can transform the fly's "halteres"—small appendages that work like tiny gyroscopic stabilizers—into a second set of normal-looking wings.

These experiments show that mutations can produce dramatic changes in the anatomical structure of organisms. And, for many evolutionary biologists, they provide a powerful confirmation of the neo-Darwinian claim that mutations provide the novel variations that natural selection needs to build new anatomical structures and animals.

Natural Selection and Mutation: Reply

Even scientists who are critical of contemporary Darwinism acknowledge that many fascinating mutations do occur. They also acknowledge that small-scale mutations do sometimes provide limited benefits to some organisms. However, critics of the mutation/selection argument doubt that such mutations can produce fundamentally new living forms.

Critics are convinced that the typical mutation examples actually demonstrate significant limits to what mutations can do. Furthermore, they doubt that mutations can produce enough of the right kind of change to produce new forms of life.[4]

Mutation Not a Source of Unlimited Novelty

For example, critics of neo-Darwinism contend that antibiotic resistance shows, at best, a limited capacity for change, not the extraordinary elasticity required by the Darwinian model. Critics of this argument say we must reexamine both examples: the enzyme defense mechanism and the mutation scenario.

Antibiotic Resistance Reply

In the case of penicillin resistance, critics agree that when penicillin is present in the bloodstream, a bacterial strain that already has a gene coding for penicillinase will have a significant survival advantage over a strain that doesn't. But critics point out that bacterial cells either have a penicillinase gene, or they don't. They do not develop such a gene when penicillin is introduced. Consequently, critics say that the enzyme defense system tells us nothing about whether mutations can produce novel forms of life. But what about the examples we have discussed where mutations do produce antibiotic resistance?

Limits to Mutation-Induced Change

Critics of neo-Darwinism acknowledge that point mutations can give bacterial cells resistance to some antibiotics. They agree that when natural selection acts upon such mutations, it can produce small-scale (microevolutionary) change. However, they do not think that mutations like those that cause antibiotic resistance can go on to produce major (macroevolutionary) changes in organisms. Instead, critics think recent discoveries about how bacteria acquire antibiotic resistance actually show that there are limits to the amount (and kind) of change that mutations can produce.

Remember one of the ways that antibiotics work. When an antibiotic binds to a critical "active site" of a target protein in an essential molecular machine, such as a polymerase or a ribosome, it impairs the bacterial cell's ability to copy or process its own genetic information. As a result, one of two things happens. Either the cell loses the ability to replicate itself, or it loses the ability to make essential proteins. Either way, without these vital functions, it's not long before the bacterium dies.

Also, remember how mutations can cause bacteria to become resistant to antibiotics. Antibiotics aim at "target" proteins in one of the bacterium's essential molecular machines. However, mutations sometimes produce a change in the shape of the active site of the target protein. Because of this change in shape, the antibiotic no longer recognizes the site as a target. And because it doesn't recognize it, it doesn't bind to it. As a result, the antibiotic no longer interferes with the machine's function. When this happens, an antibiotic-resistant strain of bacteria arises.

Mutation: A Mixed Blessing

Nevertheless, there is a downside to this process—at least, it's a downside for the resistant strain of bacteria. True, a mutation has changed the shape of the active site on the "target" protein so that the antibiotic no longer recognizes the site. On the other hand, the very same mutation usually hampers the molecular machine's ability to function—certainly not as much as the antibiotic would, but it does change a critically important protein. From the bacterium's point of view, it's a "lesser of two evils" kind of choice. Either accept a severe handicap that fools the antibiotic, or die from the antibiotic.

That's bad enough, but there's another problem. If more mutations of this type occur, they will inflict additional damage on vital systems. The cell cannot endure an unlimited number of mutation-induced changes at these critical active sites. At some point, the cell's information processing system will be damaged so badly that it stops functioning altogether. For this reason, multiple mutations at active sites inevitably do more harm than good.

Critics say this helps to explain something scientists have observed in the laboratory. Experiments show that once antibiotics are removed from the environment, the original (non-resistant) strain "out-competes" the resistant strain, which dies off within a few generations.[5]

Why does that happen? It's a classic good news/bad news scenario. A mutation gives some bacteria a resistance to antibiotics. (That's good.) However, that very same mutation also impairs that strain's ability to perform other vital functions like information processing. (That's bad.) When the environment returns to normal, the impaired mutant strain is less fit in the struggle for survival. Microbiologists refer to this as the "fitness cost" of a mutation.[6] When bacteria acquire antibiotic resistance by mutations at active sites, the new mutant strain of bacteria pays a price for its short-term competitive advantage.

And because mutations at these critical active sites come with a fitness cost, critics of neo-Darwinism argue that additional mutations of the same kind are more likely to destroy essential functions than to produce fundamentally new forms of life. This strongly suggests that there are limits to the amount of change that such mutations can produce.[7]

(But what about mutations that affect the *non*-active sites of proteins? Although these mutations do occur, they're unhelpful as agents of change. They produce one of two possible outcomes: they either have no effect whatsoever on the structure or function of the protein, or they unravel the

"A mechanism that requires a discerning human agent cannot be Darwinian. The Darwinian mechanism neither anticipates nor remembers. It gives no directions and makes no choices. What is unacceptable in evolutionary theory, what is strictly forbidden, is the appearance of a force with the power to survey time, a force that conserves a point or a property because it ***will be*** *useful. Such a force is no longer Darwinian. How would a blind force know such a thing?"*
David Berlinski, "Deniable Darwin," Commentary 101 *(June 1, 1996).*

protein and destroy its function. Either way, a mutation at a non-active site isn't likely to produce beneficial large-scale change either.)

Ready For The Big Time?

But even if it's unlikely, isn't it still at least possible that resistance-producing mutations could eventually produce a new form of life? Critics say no. In every case where mutations lead to antibiotic resistance, resistance results from small changes to a single protein molecule. For this reason, critics doubt that the kind of mutations that produce antibiotic resistance can ever produce fundamentally new forms of life—no matter how many times the same molecule is altered.

To see why, imagine that you have a television that isn't getting good reception. In fact, all you see is static. You try to improve the situation by changing just one thing: adjusting the satellite dish. After swiveling it around for a while you find a position that gives you a clear picture. You have definitely changed something—and changed it for the better. But have you fundamentally altered the system? No. You still have a television, not a microwave. And no amount of adjusting the satellite dish will turn the television into a microwave.

Critics say the same thing is true about small modifications to single protein molecules. Such changes may allow the bacterium to resist antibiotics, but no matter how many times a single protein molecule is changed by mutation, it will not transform a bacterium into a fundamentally new kind of organism, or even a fundamentally new kind of bacterium. Such mutations will only slightly modify a single protein component of a highly organized multi-part organism.

Of course, those who think that mutations can produce major biological change do not think that repeated mutations to the same protein will alone do the job. They claim that mutations to many separate proteins are necessary to produce major biological change.

Yet critics insist that mutation-induced antibiotic resistance provides no support for this claim either. They note that the mutations that cause antibiotic resistance only change a small site on the surface of a relatively large protein molecule and that these mutations do not alter the overall structure of the protein. Since the kind of mutations that produce antibiotic resistance do not change the structure of the protein components of the organism, they will not fundamentally change the organization of the organism or the organism as a whole.

And this is precisely what laboratory experiments have shown, say the critics. Bacteria reproduce very rapidly (faster than rabbits!). This allows scientists to use chemicals and other agents to induce lots of mutations in the DNA of bacteria in a short period of time. Yet, these attempts to simulate evolution over many generations have never produced a new bacterial species. As British bacteriologist Alan Linton has noted, "Throughout 150 years of the science of bacteriology, there is no evidence that one species of bacteria has changed into another."[8]

Reply to the Four-Winged Fruit Fly

It won't surprise you to learn that critics also contend that there are severe problems with the four-winged fruit fly example.

For one thing, they argue, it isn't easy to produce a fly with four wings. Skilled researchers had to induce, not one, but three separate mutations before a second set of wings appeared.[9]

In other words, researchers had to intentionally manipulate the genome of three generations of flies to produce the desired effect. The neo-Darwinian scenario says that new structures are produced by natural selection acting upon purely undirected and random mutations. Yet, the mutants that produce four-winged fruit flies survive only in a carefully controlled environment and only when skilled researchers meticulously guide their subjects through one non-functional stage after another. This carefully controlled experiment does not tell us much about what undirected mutations can produce in the wild.[10]

Meanwhile, back in the laboratory... After the third mutation, the researchers do finally produce a fly with a second set of normal-looking wings.

But critics say this reveals a huge practical problem—one that's frequently overlooked. The mutated wings don't work. Why not? The wings don't have any flight muscles. The wings are dead weight. They don't create thrust; they create drag. Not only that, these fruit flies have lost their

Mutation Scorecard

For natural selection to produce significant biological change, three conditions must be met. First, as Darwin himself said, "Nothing can be effected... unless favorable variations occur." Second, the newly arising variations must be heritable—transmitted faithfully, through reproduction. And finally, the variations must confer some advantage on the individuals that posses them. These are the three necessary conditions for natural selection. Yet, some scientists contend that neither of the two textbook examples of mutation-induced change meets all three conditions.

In the case of antibiotic resistance, the resistant bacteria have gained a temporary survival advantage. However, the variation (mutation) that produces this advantage also damages the cell's ability to copy its genetic information. For this reason such variations cannot produce major biological change.

In the case of the four-winged fruit flies, major variations have occurred and have produced a significant change in the fly's anatomy. While these mutations are heritable in the lab, they are not passed on in the wild, and are not advantageous to the fly's survival. ■

	Antibiotic-resistant Bacteria	Four-Winged Fruit Fly
Major Variation	—	+
Heritability	+	+
Advantage	+	—

stabilizers (halteres) in the process. As a result, they are severely hampered in their ability to fly.

Natural selection only favors: 1) variations 2) that produce a competitive advantage, and 3) that can be passed on to the next generation. Critics argue that the four-winged fruit fly fails at least one of these three conditions. The second set of wings is definitely a variation. But neither the four-wingers nor the transitionals that preceded them could survive in the wild. Neither can they reproduce without direct assistance from the researchers.

Rather than representing a promising opportunity for adaptive change, these flies are evolutionary dead-ends. According to the critics, the four-winged fruit fly *is* an excellent example—of a harmful and non-heritable large-scale mutation.

The Either/Or Problem: Major Mutations not Viable; Viable Mutations not Major

Critics of the mutation argument say these textbook examples point to a kind of Catch-22. Small, limited mutations (like those that produce antibiotic resistance) can be beneficial in certain environments, but they don't produce *enough* change to produce fundamentally new forms of life.[11] Major mutations *can* fundamentally alter an animal's anatomy and structure, but these mutations are always harmful or outright lethal.[12] Either way, as University of Georgia geneticist John McDonald has pointed out, the kind of mutation that natural selection requires—namely, large-scale, beneficial mutation—does *not* occur.[13] This is why critics say mutations have not turned out to be the information-rich super-variations that neo-Darwinian biologists had hoped for.

Macro Mutations?

In the 1930s, paleontologist Otto Schindewolf proposed that all the major evolutionary transformations must have occurred in single, large steps. (He proposed, for example, that at some point in evolutionary history, a reptile laid an egg from which a bird was hatched.) In 1940, geneticist Richard Goldschmidt took Schindewolf's idea one step further, suggesting that true evolutionary change takes place in the rare successes of large-scale mutations, not by the accumulation of small changes (as Darwin predicted).

Goldschmidt conceded that the vast majority of large-scale mutations produce hopelessly maladapted freaks like two-legged sheep or two-headed turtles. However, he suggested that on rare occasions a lucky accident might produce a fundamentally new form of life—an organism that was actually better suited to survive and reproduce than its "normal" siblings.

Many biologists have grave doubts about whether any such macro-mutants could survive to reproduce, noting that the large-scale mutations we know about are inevitably harmful.

Others have asked, "Even if they survived, could they reproduce? What would they mate with?" Questions like these led to the whimsical, yet appropriate, name: "hopeful monster."* Critics contend that Hopeful Monsters inevitably turn out to be hope*less* monsters. ■

The well-known antennapedia mutation causes a leg to grow on the fly's head, where the antennae should be. Unfortunately (for them), antennapedia mutants cannot survive in the wild, have trouble reproducing, and their offspring are not very robust. *Photo by © Science VU/Visuals Unlimited*

*Most modern evolutionary biologists question whether such mutation-induced changes can be passed on beyond the first generation in which they appear. Jeffrey Levinton, *Genetics, Paleontology, and Macroevolution* (Cambridge: Cambridge University Press, 1998):252-253. "As a general rule, major developmental mutants give a picture of hopeless monsters, rather than hopeful change.... The cyclops mutant of *Artemia* is lethal. The homeotic mutants of *Drosophila melanogaster* suffer similar fates."

Natural Selection and Mutation: Further Debate

Are there other mutations that might be able to produce major biological change? Evolutionary biologists say this is a live possibility. They are investigating two types of mutations they regard as promising new avenues of research: compensatory mutations, and mutations in "box genes."

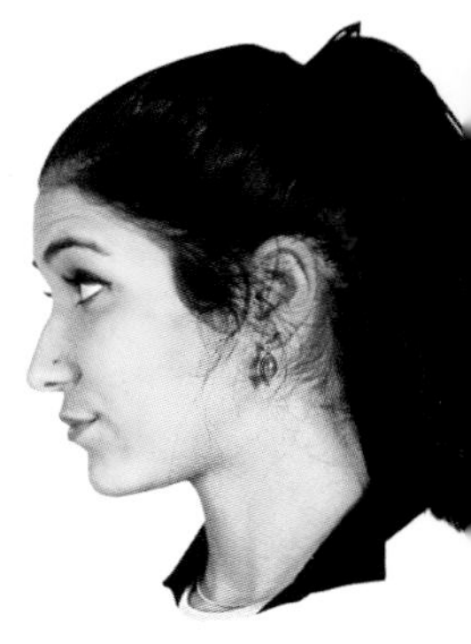

Neo-Darwinists contend that one (or both) of these mutations may have the capacity to overcome the limits that critics of neo-Darwinism have pointed out.

Compensatory Mutations

Remember our earlier discussion about bacteria that acquire antibiotic resistance as the result of a mutation. Remember also how the mutation carries a "fitness cost." Researchers have noticed that this fitness cost is sometimes offset by additional mutations.[14] Because these mutations make up for—or *compensate* for—the damage caused by the first mutation, biologists call them "compensatory mutations."

How Do They Work?

As you remember from our earlier example, the original mutation changed the shape of a "target" protein, which turned out to be one of those good news/ bad news scenarios. The good news (from the bacterium's point of view) is that the mutation gave the bacterium a resistance to a given antibiotic. The bad news is that it damaged one of the bacterium's vital systems in the process. The mutation tampered with a 'precision' component of a machine, making it less efficient because the damaged component must still interact with another precision component of the machine.

Compensatory mutations act on the "companion" protein components, and this helps restore some of the original machine function that was lost due to the first mutation. In this way, the compensatory mutations allow the bacterium to keep its resistance to antibiotics, while recovering its lost fitness. This suggests that mutations can cause new varieties to arise without loss of function to vital cellular systems.

Fitness Cost 2.0

At first glance, it appears that bacteria can become antibiotic resistant without any long-term 'fitness cost.' Appearances can be deceiving, say critics of the compensatory mutation argument.

They say that the effectiveness of the second mutation is a limited-time offer, a coupon only good for the exact environment in which it was issued. If the temperature changes in the environment,* or if the salinity changes, a whole new fitness cost comes to light. The compensatory mutation now "codes for" a protein that doesn't fold properly under the new conditions. And if it doesn't fold properly, it doesn't work properly. Or, it may not work at all.

Each time a mutation tampers with the cell's machinery, it narrows the range of conditions under which the organism can function. This makes it harder and harder for the organism to adapt to its environment. This effect increases over time, eventually forcing the organism into a macroevolutionary dead end.[15]

As a result, critics argue that additional "compensatory" mutations are no more likely to produce fundamental changes in form or structure than the original mutations were. Compensatory mutations, like the original mutations, eventually do more harm than good. Critics say this limits the potential for large-scale evolutionary change resulting from this kind of mutation.

This disagreement is far from over, and the field is wide open to more research.

Hox Mutations

Developmental biologists are investigating another kind of mutation—mutations in "hox genes"—that many neo-Darwinists think can provide a significant source of major variation in living forms.

Hox genes are "master regulator" genes that turn other genes in the cell on and off during the developmental process. Like a coach calling in plays from the sideline, hox genes determine when other genes in the cell will transmit their instructions for building proteins. Because hox genes are so important for coordinating the activities of the cell, some researchers think that mutations in these genes can cause large-scale changes in the structure of an organism. Many neo-Darwinists think that influencing the "coach" (hox genes) through mutation can provide a new source of beneficial variation.

Critics aren't so sure. They note that, precisely *because* hox genes are so influential, no experimental mutations in hox genes (so far) have proven helpful to the organism.

This makes sense, say the critics, because living organisms depend on many interrelated systems and genes. As we have seen with the four-winged fruit flies, it requires a great many coordinated changes to transform one system into another without losing function in the "in-between" steps. The more the individual parts of a system depend on each other, the harder it is to change any one part without destroying the function of the organism as a whole. Since hox genes affect so many genes and systems, it seems unlikely that they could be mutated without damaging the way some of the genes are switched 'on' or 'off.'

Will mutations in hox genes provide the source of variation that natural selection needs? Or will mutations in hox genes always harm the organism in which they occur? At this point, no one knows for sure. The debate over hox genes is taking place on the cutting edge of research, so stay tuned for further developments.

DNA Demoted?

Also on the cutting edge of research is the noteworthy claim that genes may not do as much as scientists previously thought.

By now you are well aware that the gene sequences in DNA provide the assembly instructions for building proteins. These instructions, found in the DNA of every plant and every animal, have been compared to precisely sequenced strings of code.

* This is what happens when your body "runs a temperature" when you have a cold. Your body is changing the temperature of the environment, trying to make it less hospitable to the invading bacteria or virus.

But some scientists contend that while the genes in DNA carry assembly instructions for building proteins out of amino acids, they do not carry the assembly instructions for building organs out of proteins, or for building whole creatures out of organs and other body parts.[16]

In other words, many biologists (embryologists in particular) are beginning to think that genes do not (by themselves) carry the instructions for building a whole organism or animal.[17]

Here's why they say this. Animals are made of many body parts and organs arranged in very specific ways. Body parts and organs, in turn, are made of special kinds of tissues. Each type of tissue is made of different types of cells. Each different cell type has many proteins, including some that may be unique to that particular cell type.

You can think of an animal's body as being arranged according to a kind of organizational chart. Organs and body parts are the head honchos near the top of the chart, and proteins are the working stiffs near the bottom. And that's where the problem comes in. According to many scientists, current research shows that DNA is actually closer to the bottom of the organizational ladder. Yes, it directs the building of proteins, and yes, proteins are important. But DNA does not direct how the overall body plan gets built.

Something else—some other source of information—must orchestrate the assembly of the component proteins into unique cell types, and direct the organization of cell types into various tissues and organs, and control the arrangement of organs and body parts into an overall body plan.[18]

Critics say it's a little like building a CD player. Even if you have all the information you need to build all the individual electrical components—resistors, capacitors, jumper switches, and a laser unit—you will still need additional information to arrange all the components into functioning circuit boards. You'll need even more information to coordinate the circuit boards with the mechanical components.

Remarkably, many scientists now say the same concept is true in biological systems. An

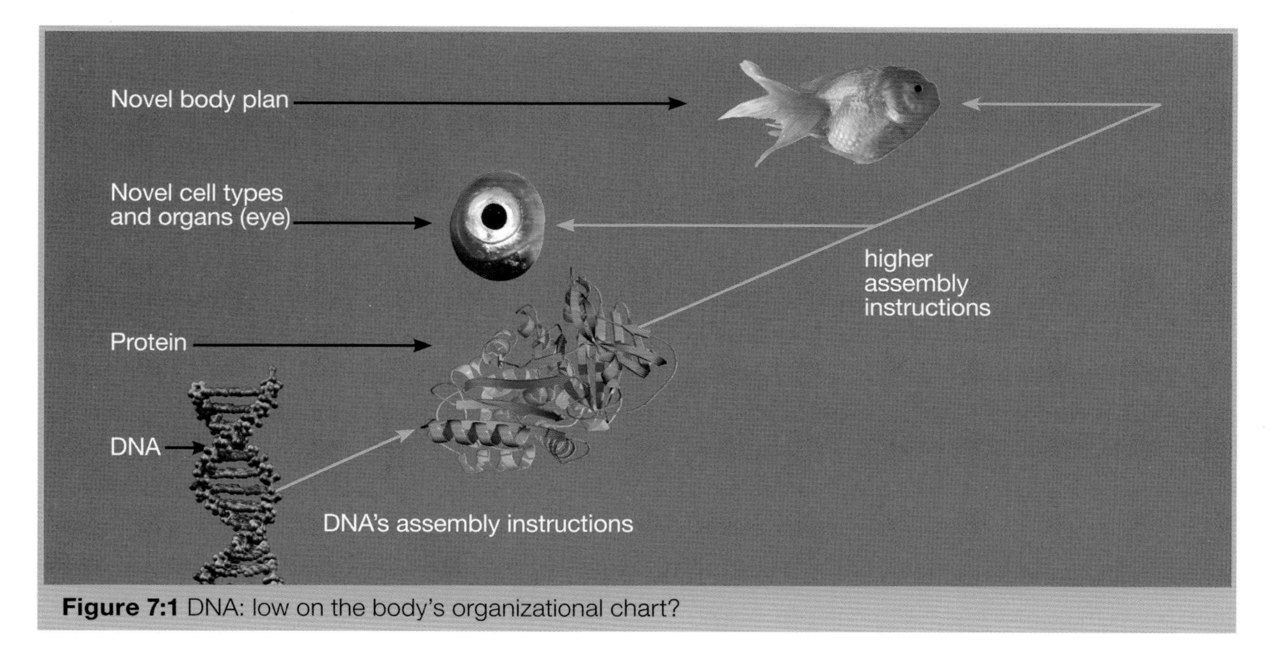

Figure 7:1 DNA: low on the body's organizational chart?

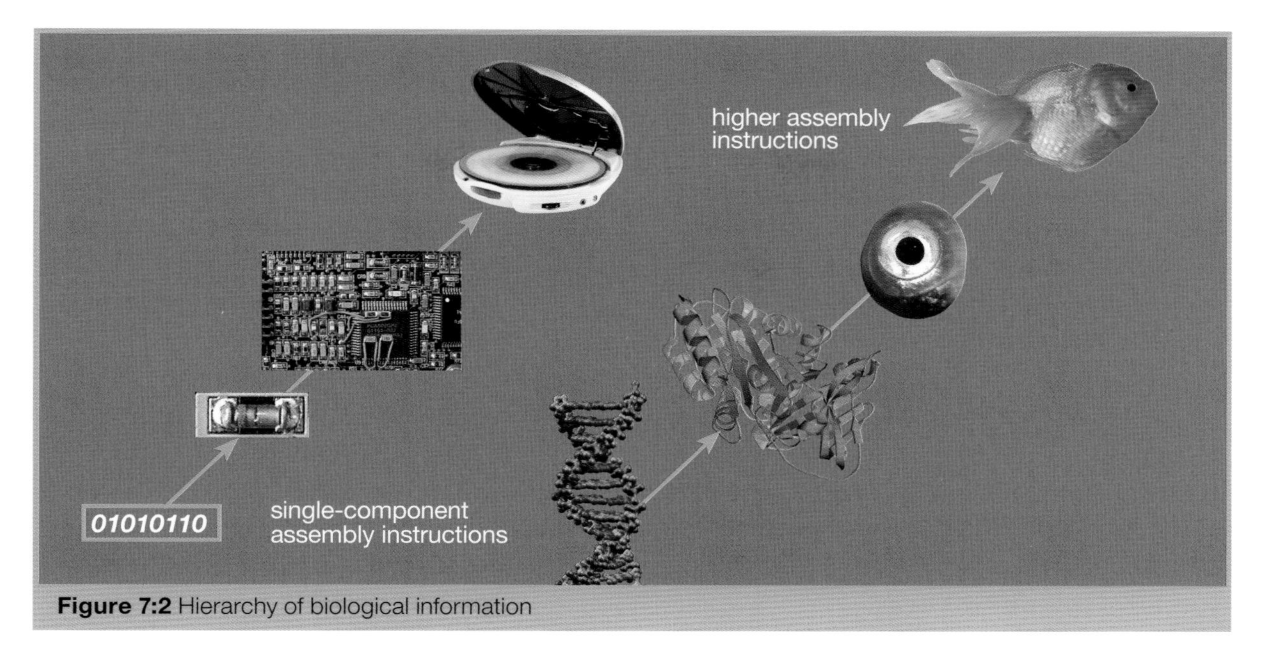

Figure 7:2 Hierarchy of biological information

organism needs genetic information to build proteins. It also needs higher-level assembly instructions to arrange tissues and organs into body plans. Scientists are not entirely sure where these higher-level assembly instructions are stored. However, these instructions are clearly necessary, and many scientists now doubt that they are stored in DNA alone.[19]

This idea poses a challenge for the neo-Darwinian scenario. According to neo-Darwinism, new biological form arises when natural selection acts on randomly occurring mutations and variations in DNA. But new research seems to say that mutations in DNA assembly instructions will produce, at best, a new protein. Higher-level instructions—for building tissues, organs, and body types—are not stored only in DNA. That means you could mutate DNA 'til the cows come home and you still wouldn't get a new body plan.[20] Even mutating hox genes won't get the job done. DNA is simply the wrong tool for the job, say many developmental biologists, and no amount of time will overcome this limitation.

Hang On A Minute

If DNA by itself doesn't control development, then what does?

This is a very hot topic, at the center of much research and debate. Some developmental biologists now think that two other cellular features—the cytoskeleton and the cell membrane—store structural information that affects how the embryo develops, but there is much that we simply don't know yet.

What we *do* know is that if DNA doesn't control development, something else must. Identifying the "something else" is one of the next great areas of research. Another question, of course, is whether the "something else" can be altered by mutation, which would provide a whole new vista of variations on which natural selection could act. Is this what the research will reveal? Or will it reveal even deeper questions?

Stay tuned.

Endnotes

1 Søren Løvtrup, "Semantics, logic, and vulgate neo-Darwinism," *Evolutionary Theory* 4:157-172.

2 Through a very interesting process—much too detailed to discuss here—some of the genes for antibiotic resistance become mobile and transferable. These mobile gene elements can even jump onto transmissible genetic elements, such as plasmids and bacterial viruses. Other bacteria in the environment occasionally encounter these mobile "jumping genes," and are able to acquire antibiotic resistance in pre-packaged form. This accounts for the rapid spread of antibiotic resistance among different bacterial genera.

3 The mutation rate for bacterial cells is about one mutated protein in one cell per 1 billion bacterial cells. On one hand, this is a very low rate of error. On the other hand, since a thimbleful of liquid can easily contain hundreds of millions—or even billions—of bacteria, a mutant bacterium can be easily generated (and detected) in an experimental setting.

4 A.D. Bradshaw, "Genostasis and the limits to evolution," *Philosophical Transactions of the Royal Society of London* B 333 (1991):289-305.

5 Henrik C. Wegener, "Ending the use of antimicrobial growth promoters is making a difference," *ASM News* 69 (2003).

6 V.D. Kutilek, D.A. Sheeter, JH Elder, and B.E. Torbett "Is resistance futile?" *Current Drug Targets - Infectious Disorders* 4 (Dec 2003):295-309.

7 Scott F. Gilbert, John M. Opitz, and Rudolf A. Raff, "Resynthesizing evolutionary and developmental biology," *Developmental Biology* 173 (1996):357-372. "Genetics might be adequate for explaining microevolution, but microevolutionary changes in gene frequency were not seen as able to turn a reptile into a mammal or to convert a fish into an amphibian. Microevolution looks at adaptations that concern the survival of the fittest, not the arrival of the fittest." They concluded: "The origin of species — Darwin's problem — remains unsolved."

8 Alan Linton, "Scant search for the maker," *Times Higher Education Supplement* (April 20, 2001):29.

9 Patricia K. Rivlin, Alice Gong, Anne M. Schneiderman, Ronald Booker, "The role of *Ultrabithorax* in the patterning of adult thoracic muscles in Drosophila melanogaster," *Dev Genes Evol* 211 (2001):55-66.

E. B. Lewis, "A gene complex controlling segmentation in Drosophila," *Nature* 276 (1978):565-570;

E. B. Lewis, "Regulation of the genes of the bithorax complex in *Drosophila," Cold Spring Harbor Symposia on Quantitative Biology* 50 (1985):155-164;

Mark Peifer & Welcome Bender, "The anterobithorax and bithorax mutations of the bithorax complex," *EMBO Journal* 5 (1986):2293-2303.

10 The survival problem is more severe than is discussed here. The first two mutated generations are crippled, unable to fly. They would be easy prey in the wild, and would probably not survive long enough to reproduce. Consequently, they wouldn't survive long enough to make it to the third mutation… even if we assume that the string of three mutations could happen consecutively in the wild.

11 Wallace Arthur, "The effect of development on the direction of evolution: toward a twenty-first century consensus," *Evolution & Development* 6 (2004):282-288.

K.S.W. Campbell and C.R. Marshall, "Rates of evolution among Paleozoic echinoderms," *in Rates of Evolution*, K.S.W. Campbell and M.F. Day, eds. (London: Allen and Unwin, 1987).

Eric H. Davidson, *Genomic Regulatory Systems: Development and Evolution* (San Diego: Academic Press, 2001).

Scott F. Gilbert, John M. Opitz, and Rudolf A. Raff, "Resynthesizing evolutionary and developmental biology," *Developmental Biology* 173 (1996):357-372.

12 Wallace Arthur, *The Origin of Animal Body Plans: A Study in Evolutionary Developmental Biology* (Cambridge: Cambridge University Press, 1997).

One exception to this rule is the **loss** of structures (e.g., eyes in cave fish or wings on island birds). In some circumstances, a species may still survive after losing features—if those features are no longer critical for living.

13 John F. McDonald, "The molecular basis of adaptation: a critical review of relevant ideas and observations," *Annual Review of Ecology and Systematics* 14 (1983):77-102.

14 Mary G. Reynolds, "Compensatory evolution in Rifampin-resistant *Escherichia coli," Genetics* 156 (December 2000):1471-1481.

15 Dan I. Andersson and Diarmaid Hughes, "Muller's ratchet decreases fitness of a DNA-based microbe," Proceedings of the National Academy of Sciences 93 (January 23, 1996):906-907.

16 Alessandro Minelli, *The Development of Animal Form: Ontogeny, Morphology, Evolution* (Cambridge: Cambridge University Press, 2003):22-24.

see also Gerd B. Müller and Stuart A. Newman, "Origination of organismal form: the forgotten cause in evolutionary theory," in G.B. Muller and S.A. Newman, eds. *Origination of Organismal Form: Beyond the Gene in Developmental and Evolutionary Biology* (Cambridge: The MIT Press, 2003):3-12.

17 Brian K. Hall, "Baupläne, phylotypic stages, and constraint: Why there are so few types of animals," *Evolutionary Biology* 29 (1996):215-253.

L. Moss, *What Genes Can't Do* (Cambridge: The MIT Press, 2004).

18 J. Sapp, *Beyond the Gene* (New York: Oxford University Press, 1987).

19 H.F. Nijhout, "Metaphors and the role of genes in development" *BioEssays* 12 (1990):441-446.

G.B. Müller and S.A. Newman, "Origination of organismal form: the forgotten cause in evolutionary theory," in G.B. Muller and S.A. Newman, eds. *Origination of Organismal Form: Beyond the Gene in Developmental and Evolutionary Biology* (Cambridge: The MIT Press, 2003):3-12.

Developmental biologist Jonathan Wells has noted, "A skin cell is different from a muscle cell, which in turn is different from a nerve cell, and so on. Yet, with a few exceptions, all these cell types contain the same genes as the fertilized egg. The presence of identical genes in cells that are radically different from each other is known as 'genomic equivalence.'" According to Wells, this equivalence presents us with a paradox. "If genes control development, and the genes in every cell are the same, why are the cells so different?" Genes, says Wells, "are being turned on or off by factors outside themselves… [C]ontrol rests with something beyond the genes…." *Icons of Evolution*, (Washington, D.C., Regnery Press, 2000):191.

20 F.M. Harold, *The Way of the Cell: Molecules, Organisms, and the Order of Life*, (New York: Oxford University Press, 2001).

21 Stephen C. Meyer, "The origin of biological information and the higher taxonomic categories," *Proceedings of the Biological Society of Washington* 117 (2004):224-225.

A New Challenge

You may have noticed that, so far in this book, we have discussed various topics in a point-counterpoint sort of fashion. You know: Darwin (or a neo-Darwinist) makes one of the classic arguments, other scientists disagree, modern scientists jump in on one side or another, and then we discuss the current state of the controversy.

As it turns out, some scientific critics of neo-Darwinism have recently gone on the offensive. They are making a new argument based upon some new discoveries about the complexity of life—structures in the cell that have many intricate and interconnected parts. Some scientists say that these structures cast doubt on the creative power of Natural Selection, because they cannot be explained by numerous, small, successive changes.

Defenders of neo-Darwinism have, in turn, critiqued this argument. We think it's important for you to know about this argument—and the controversy surrounding it. *In keeping with our point-counterpoint style, we're going to present both sides of this argument. But this time we're going to let critics of neo-Darwinism make their case first, and let the defenders of neo-Darwinism play the role of critic.*

To understand what this new debate is all about we need to examine how our understanding of the cell—the smallest unit of life—has changed in recent years. ■

Molecular Machines

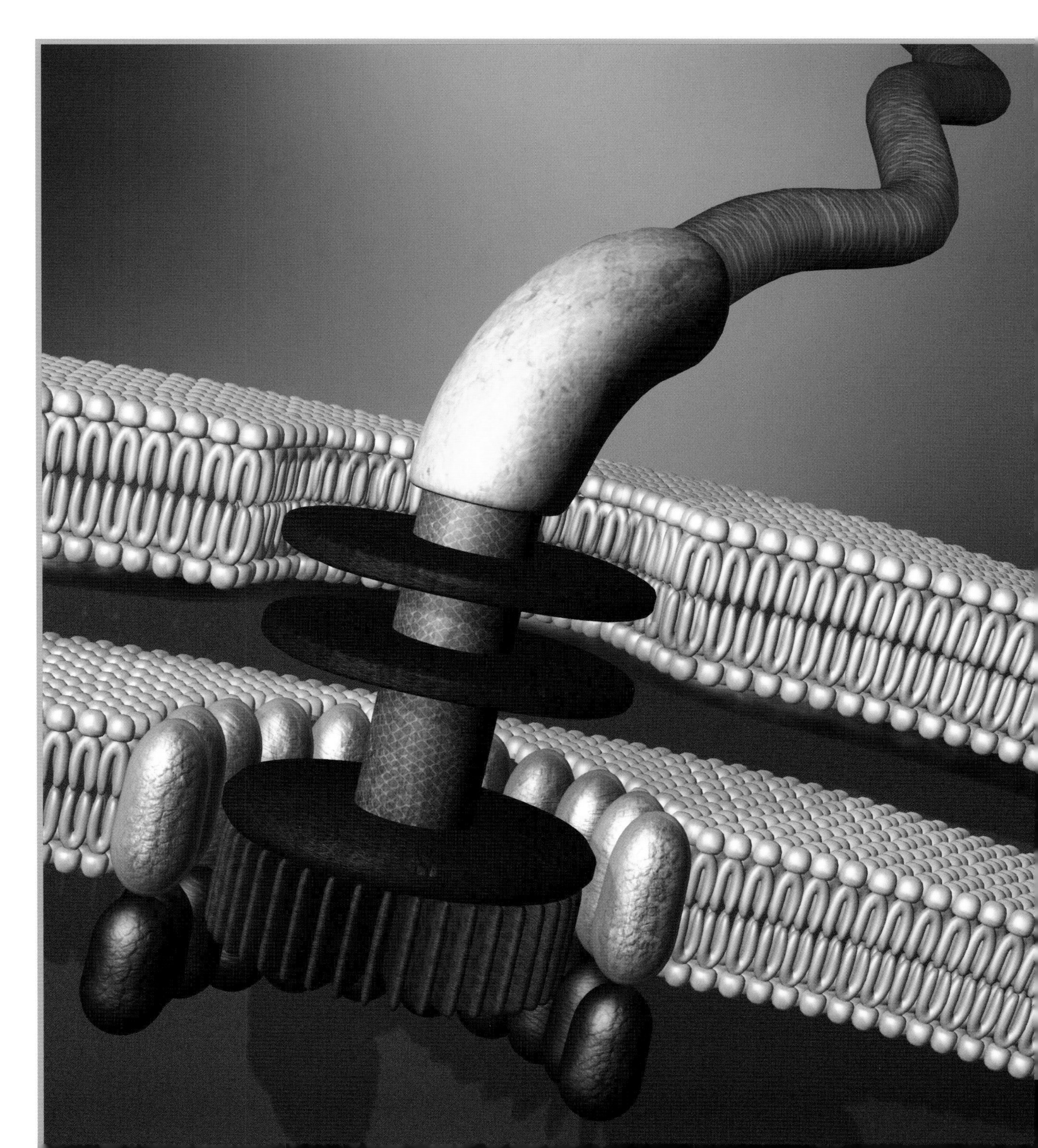

Molecular Machines: Case For

"If it could be demonstrated that any complex organ existed which could not possibly have been formed by numerous, successive, slight modifications, my theory would absolutely break down," wrote Darwin. However, he was convinced this 'absolute breakdown' would not happen, for he continues, "But I can find out no such case."[1]

Some scientists have accepted Darwin's challenge, and claim that such features—features that could not have been formed by the natural selection/mutation process—have been found.

In the late 19th century, when Darwin was working, most scientists thought the cell was a simple structure—a little blob of carbon compounds they called *protoplasm*. In contrast, thanks to the relatively new science of molecular biology, scientists have discovered many fascinating and complex structures within the cell. These include miniature motors, machines, and circuits. These complex structures are usually made of protein molecules that fit together in precise ways. Many scientists now talk about the machines in the cell as "molecular machines."[2] Bruce Alberts (former president of the National Academy of Sciences) describes what scientists now are discovering:

"We have always underestimated cells.... The entire cell can be viewed as a factory that contains an elaborate network of interlocking assembly lines, each of which is composed of a set of large protein machines.... Why do we call the large protein assemblies that underlie cell function protein machines? Precisely because, like machines invented by humans... these protein assemblies contain highly coordinated moving parts."[3]

But what do molecular machines have to do with the debate about biological origins? Quite a lot, say some scientists. For example, Michael Behe, a biochemist from Lehigh University, argues that the presence of complex machines within cells is a huge challenge to the neo-Darwinian mechanism of natural selection and mutation. To see why, let's look at one of the molecular machines that scientists have recently discovered.

Complexity and the Molecular Machine

Figure 8:1 shows a rotary motor that is found in the cell wall of certain kinds of bacteria. This motor turns a whip-like tail called a flagellum.

This spinning flagellum acts like a propeller, moving the bacterium through its liquid environment, which helps the bacteria find food.

Like other molecular machines, the bacterial motor has many components—all made of proteins. In fact, it requires 30 different protein parts for the motor's structure, 10 protein parts for the motor's sensor and control circuitry, and 10 more protein parts to construct the motor.[4]

Pictures taken with electron microscopes reveal that the bacteria's rotary motor has the same basic components as a rotary motor engineered by human beings. Both motors have a rotor, a stator, a drive shaft, O-rings, and bushings. (The rotor is the part that rotates inside the housing, while the stator is the part that remains stationary. The drive shaft connects the motor to whatever the motor is supposed to turn—like a gear, a wheel, or a propeller. A bushing is a hollow sleeve or tube that reduces the amount of wear or friction between the drive shaft—which rotates—and the housing—which doesn't.)

Motors like these have another important feature: they are complex systems. For clarity's sake, let's take a moment to spell out what "complex" means. To a scientist, a system is *complex* if it has "multiple, separate, well-matched components that interact to achieve a definite function."[5] Rotary motors in bacterial cells are certainly complex by this definition. Not only do the motors have many parts, but the parts interact in a coordinated fashion to make the motor work.

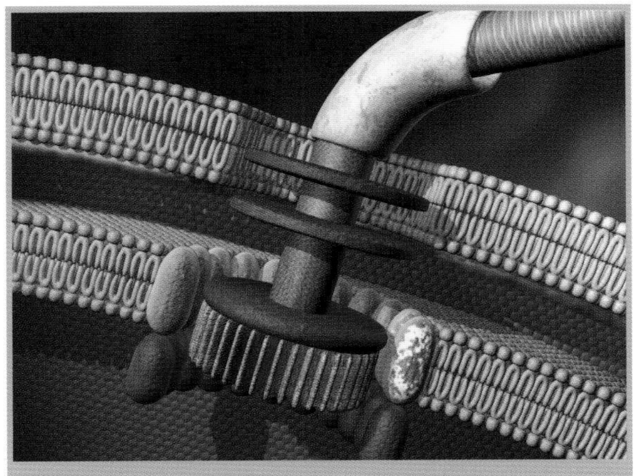

Figure 8:1 A rotary motor within a single cell
Illustration by Tim Doherty

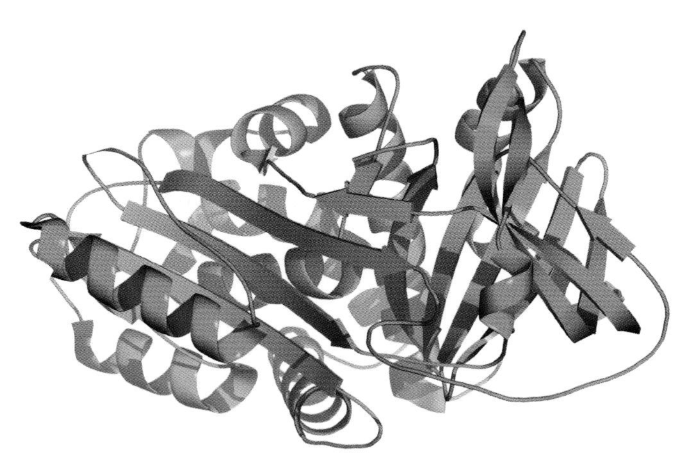

Figure 8:2 3-D model of a single protein, Heparin cofactor II (human)

Does the mechanism of natural selection acting on random mutation explain the origin of such complex molecular machines? Behe says it does not. Here's why.

Natural Selection and Complexity

Natural selection, as you already know, preserves or "selects for" structures that provide a competitive advantage. Do rotary motors confer a competitive advantage on the bacteria that have them? They certainly do. Modern bacteria that have these systems use them in their search for food. So, natural selection would definitely preserve any bacterium that developed such a rotary propulsion system.

But that's not the problem. According to scientists like Behe, the problem is not explaining why natural selection would *preserve* the motor and propeller system once it had been built. The problem is explaining how natural selection could *build* such a complex system in the first place.

Research has shown that the motor only functions after *all 30* of the motor's protein parts are in place. All 30 are required for motor function. When experimenters in the laboratory take away even one part of the motor, it stops working.

According to Darwin, complex structures can arise—but only gradually, as natural selection preserves and accumulates numerous, small variations over time. But Behe questions whether natural selection would (or could) preserve simpler, intermediate versions of the bacterial motor. Why? A motor-in-process with only 20, or even 29 of the necessary proteins does not function as a motor. And since simpler versions of the motor with fewer parts don't work, natural selection would not select or preserve such "intermediate" forms. To have a working flagellar motor that natural selection could preserve, all 30 parts have to be present together.

Because of this, Behe argues that the neo-Darwinian mechanism does not explain the origin of complex machines whose parts all have to be present before the machine will function. Behe has labeled such machines, "irreducibly complex." *(See sidebar.)* In Behe's view, natural selection and random mutation simply do not have the creative power to produce the irreducibly complex systems found in many cells.

Have modern scientists discovered the "complex organ which could not possibly have been formed by numerous, successive, slight modifications?" Many scientists think so. Others dispute this claim.

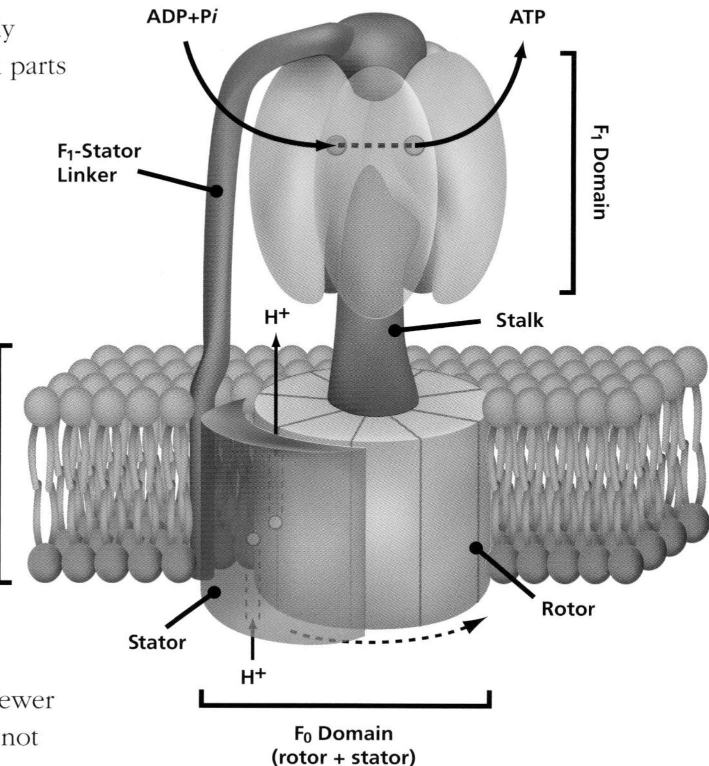

Figure 8:3 The proton-powered ATP Synthase, a turbine driven, energy producing machine in the cell.

Irreducibly Complex Systems

A system can be called irreducibly complex when it:

- Performs a specific advantageous function,
- Includes a set of well-matched, mutually interacting parts,
- Each of which is indispensable in performing the system's basic function.

Molecular Machines: Reply

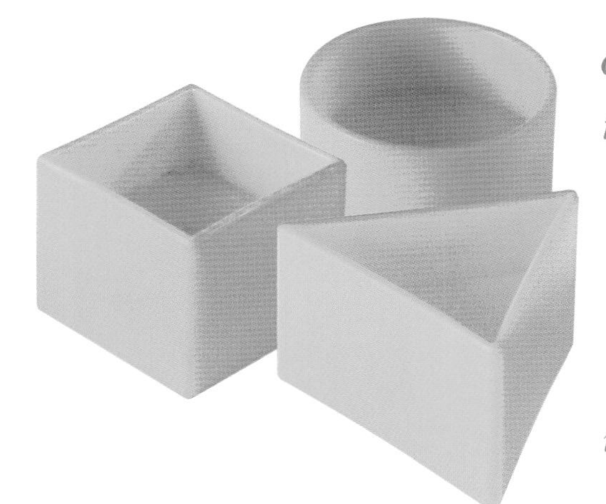

Critics of the argument from "irreducible complexity" contend that its advocates have underestimated the power of natural selection. Brown University Biologist Ken Miller and others have proposed a scenario to illustrate how complex systems could arise gradually by natural selection. They call this idea "co-option."

"I don't care what it was designed to do. I care about what it ***can*** *do."* [6]

So says Gene Kranz, NASA's Director of Flight Operations—at least, in the movie *Apollo 13*. You might recall this memorable scene in which the astronauts are in real trouble. Their spacecraft has been damaged, and they need to return to earth before their power runs out. And if that wasn't enough, they're also running out of air. The Apollo spacecraft does have "scrubber filters" that turn carbon dioxide—exhaled by the astronauts—back into breathable air. Unfortunately, the filters they've salvaged (from the damaged section of the spacecraft) are square, while the filter holders in the undamaged part of the ship are round.

Co-option

Faced with an impending disaster, engineers in the flight control center in Houston are called together. They know what they have to do.

"We have to find a way to make a square filter fit into a round receptacle, using nothing but scavenged parts the astronauts can find on board. OK, let's build a filter."

And they do. This is an example of co-option. Each co-opted part originally performed a different function in some other system. But when gathered, collected, reassembled, and recombined, the parts perform a new function—a function required to survive.

In a similar way, advocates of the co-option hypothesis think that natural selection could, in principle, build complex structures piece by piece. It could do so by co-opting portions of simpler systems—systems that performed different functions earlier in their evolutionary history. Natural selection might have "borrowed" components from one molecular machine and "tinkered" with other components from another molecular machine. According to this scenario, natural selection can build complex machines by

Figure 8:3 Apollo 13 and its co-opted filter
Photo courtesy of NASA

preserving simpler structures and *combinations* of those structures—provided each one performs some advantageous (and therefore, selectable) function at each step along the way.

Advocates of the co-option hypothesis acknowledge that if parts are removed from the molecular machines we see *today*, they will often lose their *present* function. However, they point out that this is not the same as saying that machines with fewer parts would lose *all* function.

Nor does it mean that there couldn't have been a simpler ancestral structure that performed *some* function. (Not the same function as it has today, perhaps, but some useful function.) And because it *was* useful, natural selection would have preserved it. Thus, it would be available as raw material later, ready to be combined to make an altogether new structure.

The Co-Opted Motor

And that is how critics of the irreducible complexity argument think the bacterial flagellar motor may have emerged. Brown University biologist Ken Miller, for example, has noted that many of the protein parts we see in the flagellar motor are also found in a simpler system—a machine that pumps proteins through the cell wall. *Figure 8:4* shows one of these systems, called a needle-nosed cellular pump or a "type III secretory system." This pump includes 10 proteins that are also found in the rotary motor. Miller says this shows that protein parts from the flagellar motor can perform other functions—functions that natural selection could have preserved in simpler structures. Such structures might have been co-opted along with other simpler parts to form the motor assembly for the bacterial flagellum.

Since the proteins in the pump already performed some function, natural selection could have preserved them. Once borrowed for the motor assembly, they could perform another function—and could once again be favored by natural selection. In this way, the bacterial motor could have developed from protein parts that already existed, and that were being used by earlier, simpler machines. So, co-option advocates conclude, the flagellar motor only *appears* to be a "complex organ which could not possibly have been formed by numerous, successive, slight modifications." Instead, through the process of co-option, natural selection and mutation could have assembled a complex molecular machine from a patchwork of co-opted machine parts made of proteins.

Figure 8:4 The Needle-nosed cellular pump

MOLECULAR MACHINES: FURTHER DEBATE

Some scientists are critical of the co-option argument. They note that it is easy to talk in vague generalizations about how complex machines arise when natural selection borrows some existing (simpler) parts from other organisms. But providing a detailed explanation of how a particular structure arose in this way is another matter entirely. In the case of the bacterial flagellar motor critics see several formidable obstacles.

Co-option's critics point out that simply calling a complex system "a patchwork of co-opted protein pieces" glosses over many major difficulties involved in building a bacterial flagellar motor.

For one thing, the co-option hypothesis underestimates the problem of coordinating the necessary transformations. How does an undirected process incorporate the "borrowed" parts to form a new system—without disrupting the old system? What natural process ensures that previously unrelated parts fit together well enough to form a new, functioning system? [7]

Critics, including many neo-Darwinian biologists, contend that these questions remain unanswered, and are skeptical of the co-option hypothesis. As H. Allen Orr has said, "You may as well hope that half your car's transmission will suddenly help out in the airbag department."[8]

But there are other difficulties.

Biologists have analyzed the genes and proteins found in both the cellular pump and the bacterial motor, and have found that the proteins in the motor are older than those in the pump.[9] The motor came first and the pump came later. Thus, if anything, the pump evolved from the motor, not the reverse.

Furthermore, critics of co-option point out that the bacterial motor is a machine with about 30 structural parts. While roughly 10 of these protein parts are found in the needle-nose pump, the other 20 are found in no other known bacteria or organism. They are unique. So, where are you going to 'borrow' them from? they ask.[10]

A Deeper Problem

Critics of co-option say that ongoing research has revealed another objection to the co-option scenario. They argue that even if all the genes for making the individual parts could be brought together in a single bacterium by natural selection, these parts would still have to be assembled in the correct sequence. Building the molecular motor requires genetic assembly instructions which not only tell the cell *how* to

Assembly Line... Protein Style

The molecular motor is built in a precise sequence, following the genetic assembly instructions. Working from the inside out, one mechanism counts the number of components in the ring structure of the stator. Once that's assembled, a feedback mechanism stops work on that piece. Then, a rod is added, followed by a ring. Then, another rod is added. Then comes the U-joint. Once the U-joint grows to a certain size, and is bent at the proper angle, another feedback mechanism shuts off work on that part. Then, the assembly line fabricates components for the propeller. ■

build the specific protein parts, they direct *when* these parts will be built—the precisely timed sequence of how the pieces will be assembled into a working machine.

What's more, these genetic assembly instructions are both "read" and processed in the cell—by *another* network of protein machines. And that's another problem, say co-option's critics: the system of proteins that reads and processes the genetic assembly instructions is, itself, irreducibly complex. Remove a protein machine from this system, and motor construction shuts down.

In other words, even assuming the presence of all the necessary genes and protein parts, the only way co-option can explain the origin of one irreducibly complex system (the bacterial motor) is by assuming the pre-existence of another irreducibly complex system (the system of protein machines that reads and processes genetic information).[11] Critics of co-option say this is rather like explaining the origin of machines by saying that a machine that makes machines made them.

Where Do We Stand?

Critics of the co-option hypothesis argue that it glosses over many practical difficulties involved in constructing complex molecular machines. On the other hand, critics of the argument from irreducible complexity say, "It is not valid... to assume that, because [Behe] cannot imagine such pathways, they could not have existed... We face not only an absence of data, but also the awful fact that we ourselves are evolved creatures with limited cognition and imagination."[12]

Is the problem simply our lack of imagination? Or, does the problem rest with natural selection's lack of creative power? Will scientists find an adequate, mechanistic explanation for irreducibly complex systems? Have they already? Or will researchers conclude that the neo-Darwinian mechanism ultimately comes up short? These are some of the most lively scientific questions of our time.[13] It is our hope that some of you will be among the researchers that will shed more light on these questions in years to come. Or, maybe you'll help discover some bigger questions.

Where will the evidence lead?

ENDNOTES

1 Charles Darwin, *On the Origin of Species* (Cambridge, Mass: Harvard University Press, 1964 [Facsimile of the First Edition, 1859]) Chapter 6: "Difficulties of the Theory":189.

2 Sharon A. Endow, "Kinesin motors as molecular machines," *BioEssays* 25 (2003):1212-1219.

3 Bruce Alberts, "The cell as a collection of protein machines: preparing the next generation of molecular biologists," *Cell* 92 (February 8, 1998):291-294.

4 When we say that the cell itself constructs its own motor, this is not as easy as it sounds. The cell itself builds [the flagellar motor] from the inside out. It must make each component at the right time, in the right number, and in the right order. The cell also has its own feedback loop to tell it if the motor has been assembled properly.

5 Michael J. Behe, *Darwin's Black Box: the Biochemical Challenge to Evolution* (New York: The Free Press, 1996):39.

William A. Dembski, *No Free Lunch* (Lanham, MD: Rowman & Littlefield Publishers, 2002):285.

6 Quoted from the Universal/Imagine Entertainment movie Apollo 13, Dir. Ron Howard, based on *Lost Moon* by Jim Lovell & Jeffrey Kluger, (Universal City, CA, Universal Home Video, 1995), videocassette (VHS) (140 min.).

7 Michael J. Behe, David W. Snoke, "Simulating evolution by gene duplication of protein features that require multiple amino acid residues," *Protein Science* 13 (2004):2651-2664.

8 H. Allen Orr, "Darwin v. Intelligent Design (Again)," *Boston Review* (December/January 1996-1997):29.

9 Scott A. Minnich and Stephen C. Meyer, "Genetic analysis of coordinate flagellar and type III regulatory circuits in pathogenic bacteria," in *Design and Nature II: Comparing Design in Nature with Science and Engineering*, eds. M.W. Collins and C.A. Brebbia (Southampton: WITpress, 2004):295-304.

Milton H. Saier, "Evolution of bacterial type III protein secretion systems." *Trends in Microbiology* 12 (2004):113-115.

10 Scott A. Minnich and Stephen C. Meyer, *op.cit.*

11 Scott A. Minnich and Stephen C. Meyer, *op.cit.*

12 Jerry Coyne, *Nature* (Sept 19, 1996):227-228.

13 Kenneth Miller, "The Flagellum Unspun"; Michael Behe, "Irreducible Complexity: Obstacle to Darwinian Evolution," in *Debating Design*, eds. Dembski and Ruse, (Cambridge University Press, 2004).

Special Studies

Natural Selection
as Survival of the Fittest

What Fossils
Can't Tell You

Special Study: Natural Selection as Survival of the Fittest

Just about everyone has heard the popular description of natural selection as "survival of the fittest," a term Darwin credited to Herbert Spencer. The grassroots appeal of this catch-phrase was very great, but it sometimes concealed some sloppy reasoning.

For example, someone might say that a particular organism survived because it was more "fit" than its competitors. Unfortunately, when we examine this "explanation" closely, we find that it only takes the very fact it was supposed to explain and says it in a different way. It would be similar to saying, "Amy and Margaret disagree about this issue because they have a difference of opinion."

Now, let's look at an example from biology. Let's say we have a population of 20 finches making up 10 breeding pairs. We want to predict which of these pairs will leave the most offspring in the next generation. To be a good example of natural selection, our prediction should have something to do with a heritable trait that is useful or profitable to the finch that possesses it. Such a trait would help the finch survive, making it more likely to reproduce.

Our first hypothesis is that body weight is a good measure of health and vitality, so we predict that the finches with the greatest body weights should produce more offspring, making them the "fittest" within that population. We observe the population for several generations and discover that there is no apparent relationship between body weight of the parents and the number of surviving offspring.

Discouraged by this observation, now we hypothesize that the fittest finches are hearty, a trait that can be measured by the ability to produce and sustain large numbers of eggs. To test the new hypothesis, we settle down to watch. Sure enough, we see that some finches produce and hatch more eggs than others.

But what does this tell us? It certainly doesn't tell us *why*. All we can say now is that some finches leave more offspring (our definition of "survival") because they produce and sustain more eggs (our definition of "fittest"). Cause and effect have flowed into each other, which makes the reasoning circular.

Many philosophers of science have studied this problem.* Some think that biologists can avoid such circularity by carefully and specifically identifying the trait that is responsible for the competitive advantage and reproductive success of the population being studied. Some scientists are unsure, and some think the concept may be incurably circular. ■

* Ronald H. Brady, "Natural selection and the criteria by which a theory is judged," *Systematic Zoology* 28 (1979): 600-621.

Special Study: What Fossils Can't Tell You

Remember our earlier discussion of the fossil evidence, and what various scientists say it tells us about the history of life. Advocates of Common Descent argue that the fossil record shows a general trend from simple to complex forms of life. To these scientists, this evidence suggests that more complex forms of life evolved from earlier simple ones.

Remember also that because the fossil record does show evidence of at least some transitional forms (like *Archaeopteryx* and the mammal-like reptiles), advocates say it is reasonable to think that others will be found.

One group of critics, as you recall, pointed out that when we study the fossils we have actually found, the evidence does not lead us to connect the separate lines of descent into a single branching tree. These scientists see the picture of life as a pattern of disconnected, parallel lines—an orchard with separate trees.

Scientists have another criticism of the fossil argument, but since it wasn't based on the fossils *per se*, we had to leave this criticism until now. These critics of Common Descent say it's time to take a closer look at what it would really take to transform a reptile into a mammal or a bird. They question whether the transitions envisioned by advocates of Common Descent are really plausible.[1]

Two Transformations: Reptile-to-Mammal/Reptile-to-Bird

Both mammals and birds are supposed to have descended from reptiles. (Mammals are thought to have descended directly from reptiles, but there is a vigorous debate about whether birds evolved directly from reptiles or from theropod dinosaurs which, in turn, evolved from reptiles.) Either way, the differences we see in the organs and anatomy of reptiles and their supposed descendents raise some difficult questions. How do new anatomical structures arise from old ones, for example?

Degree of Difficulty

Some scientists question whether an undirected process like natural selection acting on random mutations could coordinate all the changes to organ systems that would be needed to convert reptile organ systems into new and distinctive organ systems for mammals and birds. Of course, other scientists disagree. In essence, the two sides disagree about the "degree of difficulty" involved in such transformations.

In biology, an "easy" transition doesn't mean simple or rapid. It means that the transition between two systems—going from System A to System B—could occur: 1) by known mechanisms, and 2) without losing function anywhere between A and B. But are critical evolutionary transitions really "easy" in this sense? Let's explore this further, beginning with the reptile-to-mammal transformation.

Systems, Not Just Skeletons

The skeletons of mammals and reptiles do have some similarities, and it's fairly easy to envision a series of intermediates between them. As we've seen, some paleontologists think they have found such intermediates: the mammal-like reptiles.

But critics say that the superficial resemblance between skeletons is not all there is to the story. Transforming a reptile into a mammal would involve more than simply changing some bones along the way. It would also involve major changes in organs and organ systems. Let's look at some examples.

Reptiles and mammals reproduce very differently. Most reptiles lay eggs, while mammals carry fertilized eggs internally in a placenta and bear live young. When reptile eggs hatch, the young must immediately fend and forage for themselves. When mammals are born, the mother nourishes the young by lactation.[2]

Another example is that reptiles are cold-blooded, which means that body temperature rises and falls with the temperature of the environment. Yet, mammals are warm-blooded, which means they can regulate their body temperature internally.

Reptiles and mammals also have different vascular and circulatory systems. Both reptiles and mammals have dual circulation, which means there are two pumping systems running simultaneously. The *venous system* pumps deoxygenated ("spent") blood into the heart. The *arterial system* pumps the fresh blood out to the body. However, most reptiles have a three-chambered heart while mammals have a four-chambered heart. Let's take a quick look at how they work.

In a three-chambered heart, one chamber (called an *atrium*) receives deoxygenated blood from the venous system, while the other atrium receives oxygenated blood from the lungs. Both of these chambers empty into the third chamber, a single *ventricle*. From a practical standpoint, this means that the "spent" blood and the "fresh" blood get mixed together to some extent before getting pumped out to the body.

A four-chambered heart works differently. In a four-chambered heart, two chambers (one atrium, one ventricle) pump deoxygenated blood from the venous system to the lungs, where it is "recharged" with oxygen. The heart's other atrium-ventricle combination pumps the oxygenated blood into the arterial system which carries the fresh blood out to the body. Because the three- and four-chambered hearts work differently, it's not surprising that they have some different parts, and a different overall arrangement of those parts.

Transformations Made Easy?

Even so, is it possible that the organ systems we've just listed are similar enough that a series of small and undirected modifications could produce the differences we see in reptiles and mammals? Some scientists say yes, but others doubt this.

Why? Well, first, critics point out that any transition from a reptile to a mammal would require the development of completely new organ systems. Transforming the reproductive system, for example, is not just a question of changing where the eggs grow. It also requires the development of completely new organs like the placenta and mammary glands. As we have seen before, some scientists doubt that natural selection and random mutations have this kind of creative power.

Doubters of the reptile-to-mammal transition also argue that many necessary anatomical changes would have to take place in a coordinated fashion. Transforming a reptile to a mammal requires the step-by-(survivable) step conversion of many, separate physiological systems. It requires a coordinated change of the respiratory, circulatory, and reproductive systems, to name just a few.[3] And all of the intermediate ("in-between") organ systems must work, and in many cases, they must work *together*. Since vital organs are—well, *vital*—even a temporary loss of organ function could result in the death of the transitional animal forms. And remember: dead animals don't reproduce, and animals that don't reproduce can't evolve.

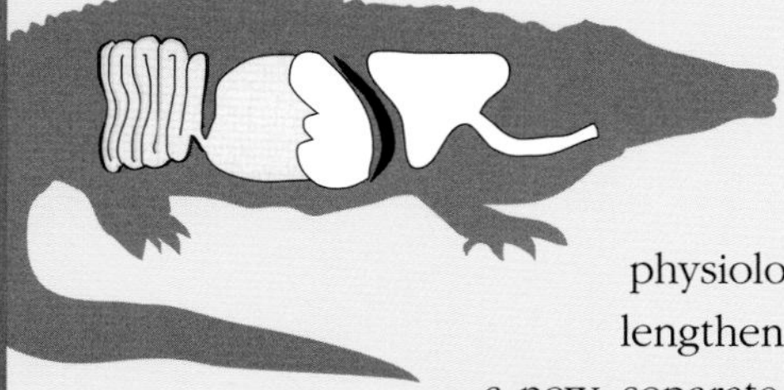

For example, any transition from a three-chambered heart to a four-chambered heart requires a series of coordinated physiological and anatomical changes, including: 1) lengthening and attaching the existing septum to create a new, separate ventricle chamber, 2) replacing the forked abdominal aorta and two aortic arches with a single aorta, 3) rerouting the pulmonary arteries and veins, and 4) making various secondary structural changes to the walls and valves between chambers.

The need to change so many anatomical features and still maintain function at every step along the way raises some difficult questions about the viability of a series of transitional stages between a three- and four-chambered heart. Would the new chamber arise before the new veins, arteries, and septa? Would it arise before or after the new, single aorta emerged and the new plumbing was rerouted to support it? If the new chamber arose before these other features, how would it work? If it arose after the other features, what are the odds that all these new structures—veins, arteries, septa, and aorta—would fit together properly and function in concert with the new chamber?

Big Steps or Little Steps?

Such questions have led some evolutionary biologists to doubt that an organ system like the heart could maintain its function while undergoing a series of small mutational changes in such a large number of critical anatomical features. Instead, some evolutionary developmental biologists now think that systems like the four-chambered heart must have arisen *as complete systems* as the result of mutations that occurred early in the development of the organisms in question.[4] They argue that because such mutations occur so early in development, they have the potential to produce larger-scale change and to affect the system as a whole, rather that just individuals parts of the organ or system. Such mutations would therefore eliminate the need to coordinate the many minor modifications required within the classical neo-Darwinian model.

Critics of the evolutionary developmental model agree with this critique of neo-Darwinism, but question the power of early-acting mutations to produce beneficial large-scale change.[5] They point out that though some mutations do occur early in the development of organisms, these mutations inevitably disrupt the orderly processes of organ system construction. Experimental results to date show that such mutations either destroy the organism altogether or cause it to lose pre-existing anatomical structures. Either way, this kind of mutation does not produce beneficial large-scale change.[6] *{cont. on page 133}*

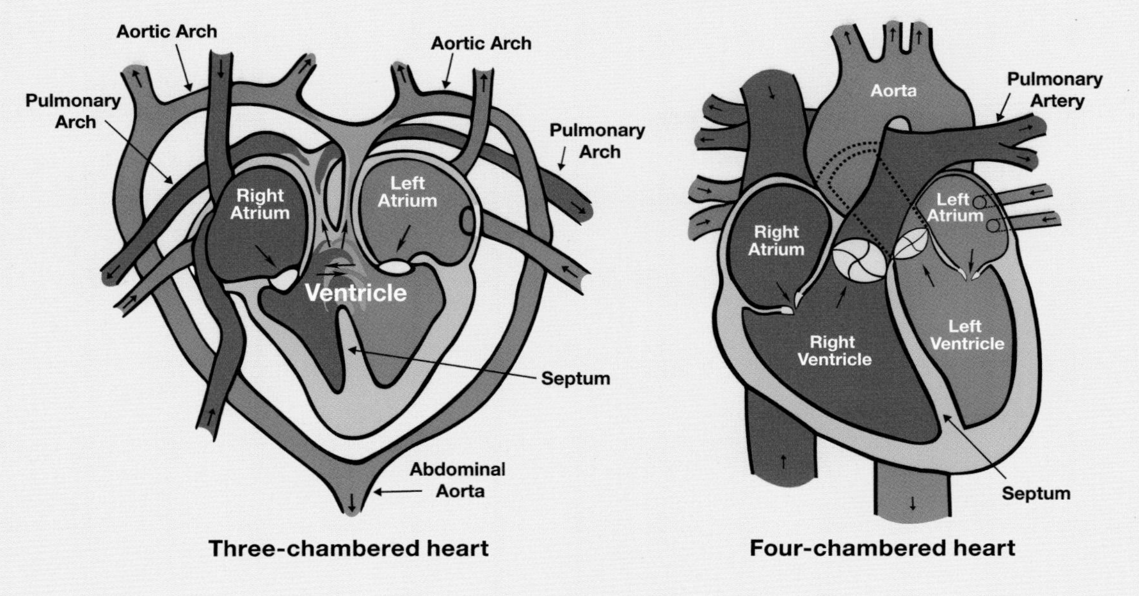

Three-chambered heart

Four-chambered heart

A Heart-to-Heart Connection?

Some reptiles—the "crocodilians"—(crocodiles, alligators, caimans, and gavials) do have four-chambered hearts. Unfortunately, this does not solve the problem of how the four-chambered heart first arose in mammals. There are three reasons for this. First, the crocodilians are not among the mammal-like reptiles that are thought to be ancestral to the first mammals. Second, even if crocodiles *were* ancestral to the mammals, we'd still need to explain how the crocodile's four-chambered heart first arose from the three-chambered heart of the typical reptile. In other words, we'd still have to solve the same evolutionary puzzle of how a four-chambered heart could evolve step-by-step from a three-chambered heart while avoiding a non-functional (lethal) intermediate form. Third, though the crocodile heart is a four-chambered heart, it is a special kind of four-chambered heart*—as different from the mammal's four-chambered heart as it is from the reptile's three-chambered heart. Here's what we mean.

The mammalian heart uses a passive one-way valve that opens under pressure, which allows blood to flow through. Unlike the mammalian four-chambered heart, the crocodilian heart has an active control mechanism that enables the croc to redirect (technically, *shunt*) blood back through its tissues when it is relaxed and in no danger. This allows the croc to recycle partially oxygenated blood for use in the body when the animal is resting. To do this, the crocodilian heart employs a cogged-tooth valve that opens and closes in response to the level of adrenaline in the crocodile's system. This active control system allows the croc to channel blood to the lungs to recharge with oxygen when the croc needs to fight or flee.** This crocodilian anatomical system (and several of its components—especially the cogged tooth valves) is unique. As Craig E. Franklin, an evolutionary geneticist from the University of Queensland has explained, "What we've shown here is an evolutionary novelty. The crocodile heart is the most complex and most bizarre in terms of its plumbing."***

---

*Craig E. Franklin and Michael Axelsson, "The intrinsic properties of an *in situ* perfused crocodile heart," *Journal of Experimental Biology* 186 (1994):269-288.

see also Michael Axelsson, "The crocodilian heart: more controlled than we thought?" *Experimental Physiology* (2001):785-789.

**Craig Franklin, Frank Seebacher, Gordon C. Grigg, and Michael Axelsson, "At the crocodilian heart of the matter," *Science* 289 (September 8, 2000):1687-1688.

***S. Milius, "Toothy valves control crocodile hearts," *Science News 2000* 158 (August 26, 2000):133.

Reptile to Bird: Survival of the Transitionals

Scientists have also raised questions about the difficulty of transforming reptiles into birds.[7] Specifically, they question whether it is possible to transform the reptilian respiratory system into an avian respiratory system. Here's why.

Reptiles, from which birds are thought to have descended, have a diaphragm breathing system. (Mammals use a diaphragm system, too.) The avian (bird) breathing system is totally different. Let's compare the two.

How Do Animals with Diaphragms Breathe?

For animals that have a diaphragm, breathing is a two-stage process. Stage one, air goes into the lung; stage two, air goes out of the lung—two directions. Reptile and mammal lungs are made up of millions of tiny balloons, called *alveoli,* which expand and contract as the animal breathes. But the lungs have no muscles, so how do they inflate and deflate? For mammals and some reptiles—including those believed to be the ancestors of birds—the lungs sit in an airtight sac or chamber. The bottom of this chamber is sealed with a large, thin muscle called the diaphragm. The dome-shaped diaphragm completely separates the chest cavity from the abdomen. When the breathing muscles force the diaphragm down, it changes the air pressure in the chest cavity, and the lungs fill with air. (To get a better idea of how the diaphragm breathing system works, try making the do-it-yourself lung on the next page.)

How Do Birds Breathe?

No known bird has a diaphragm, and a bird's lungs do not change size when it breathes. Instead, a bird has a network of air sacs that work together like bellows, drawing air in and through the bird's lungs, much like a furnace draws air in and through the ductwork of a house. Air sacs are unique to birds.

A bird has two sets of air sacs, the posterior and the anterior. Birds inhale and exhale by moving their ribs inward and outward, and by raising and lowering their very large sternum. (You already know about the bird's sternum—except you probably call it the wishbone.)

Also unique to birds is the way their lungs function. Unlike reptile lungs, bird lungs have an opening at each end. Air flows *through* the bird's respiratory system—a one-way path, which provides an almost continuous supply of fresh, oxygen-rich air in the lungs. Also, in contrast to a reptile's two-stage breathing cycle (inhale-exhale), birds require four breathing stages to circulate air through their respiratory tracts (inhale-exhale-inhale-exhale). *{cont. on page 135}*

Do-It-Yourself Lung: Part 1

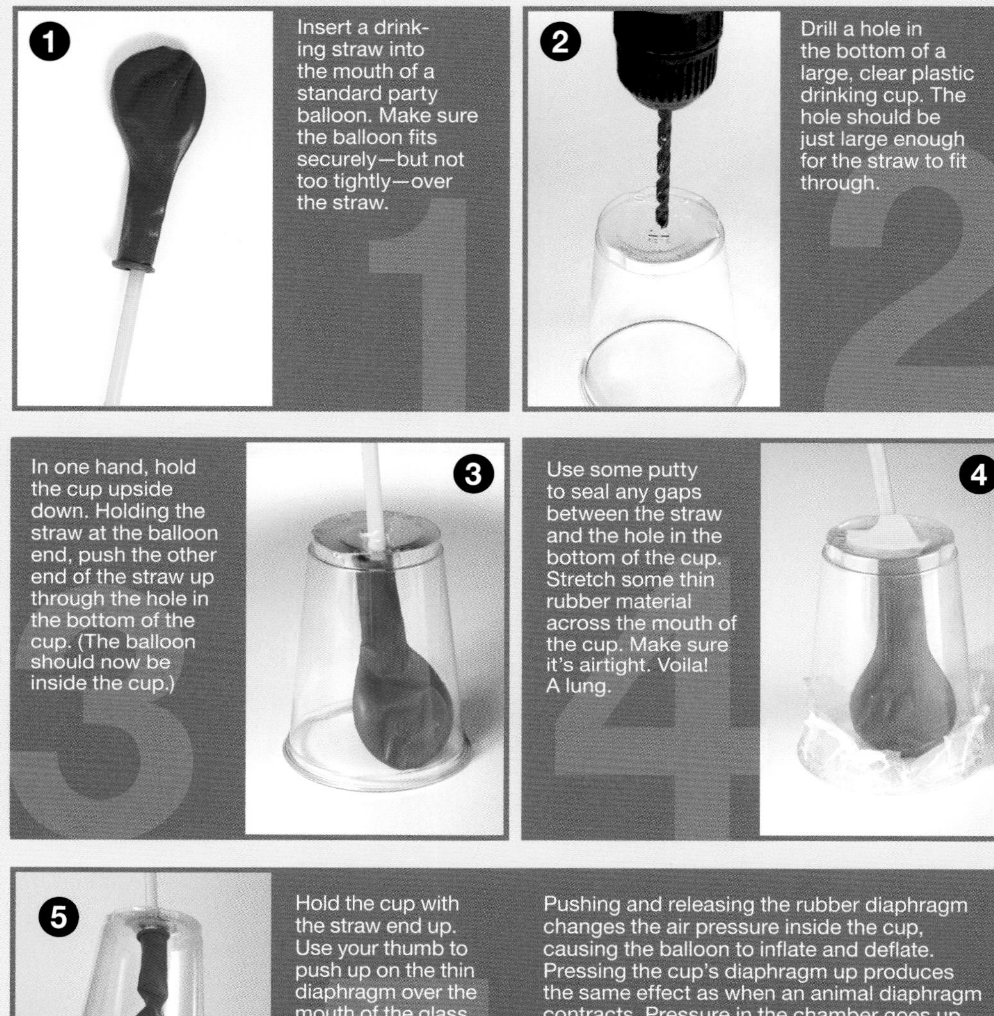

1 Insert a drinking straw into the mouth of a standard party balloon. Make sure the balloon fits securely—but not too tightly—over the straw.

2 Drill a hole in the bottom of a large, clear plastic drinking cup. The hole should be just large enough for the straw to fit through.

3 In one hand, hold the cup upside down. Holding the straw at the balloon end, push the other end of the straw up through the hole in the bottom of the cup. (The balloon should now be inside the cup.)

4 Use some putty to seal any gaps between the straw and the hole in the bottom of the cup. Stretch some thin rubber material across the mouth of the cup. Make sure it's airtight. Voila! A lung.

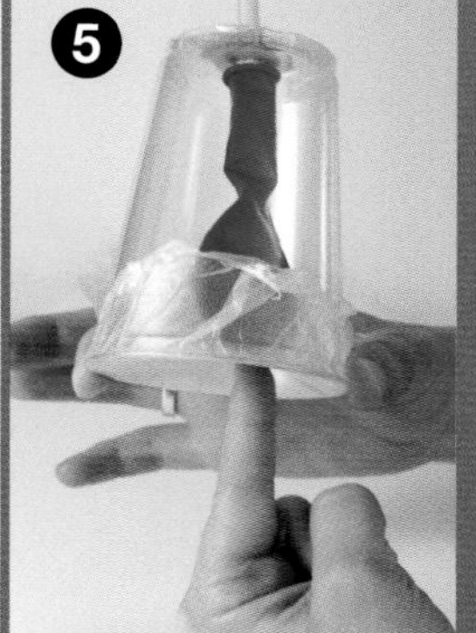

5 Hold the cup with the straw end up. Use your thumb to push up on the thin diaphragm over the mouth of the glass. What happens? The balloon deflates. You have exhaled. Remove your thumb and let the rubber sheet spring back. What happens? The balloon inflates. You have inhaled.

Pushing and releasing the rubber diaphragm changes the air pressure inside the cup, causing the balloon to inflate and deflate. Pressing the cup's diaphragm up produces the same effect as when an animal diaphragm contracts. Pressure in the chamber goes up, and the air in the balloon (lung) is forced out. When the diaphragm is released, pressure in the chamber decreases, and air flows in to equalize the pressure.

The diaphragm may be thin, but it's quite powerful. You know this if you've ever had the hiccups. A hiccup is breathing in, unintentionally. The diaphragm contracts involuntarily—rather like a muscle spasm. This pulls air into the lungs very quickly. The opening between the vocal cords closes unexpectedly, and the vocal cords come together so quickly they vibrate, causing the familiar 'hic' sound.

Here's a brief overview of how the bird breathing cycle works:

- In ***stage 1*** (first inhalation), some of the fresh air goes into the lungs, but most of the it goes into the posterior air sacs.
- In ***stage 2*** (first exhalation), the air moves from the posterior air sacs into the lungs, where oxygen is exchanged for carbon dioxide.
- In ***stage 3*** (second inhalation), the air moves from the lungs to the anterior air sacs.
- In ***stage 4*** (second exhalation), the air passes up the bronchus and out of the bird's system.

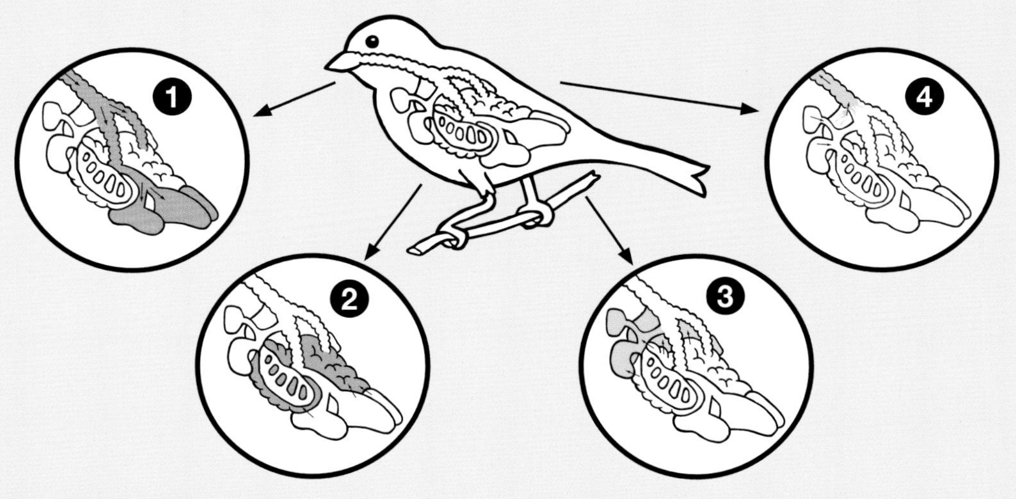

The Plot Thickens

Here are two critical questions. What would it take to transform a reptile's diaphragm breathing system into the birds' flow-through system? And, could this occur in a series of step-by-(survivable)-step transformations?

The two stage breathing apparatus of reptiles (and mammals) is an integrated system, made up of separate parts working together to perform a function. The four stage flow-through breathing apparatus of birds is also a functionally integrated system that performs the same function. But these two systems are very different. What's more, they are different *as systems*—that is, they perform the same function in different ways using different anatomical structures in different configurations. From an evolutionary point of view, this poses a problem.

A gradual evolutionary scenario from reptilian respiration to avian respiration requires a series of intermediate structures and systems. Each of these intermediates would have to be advantageous to the organism, and they would all have to be fully functional at every step along the way.

The Four Stage Avian Breathing System

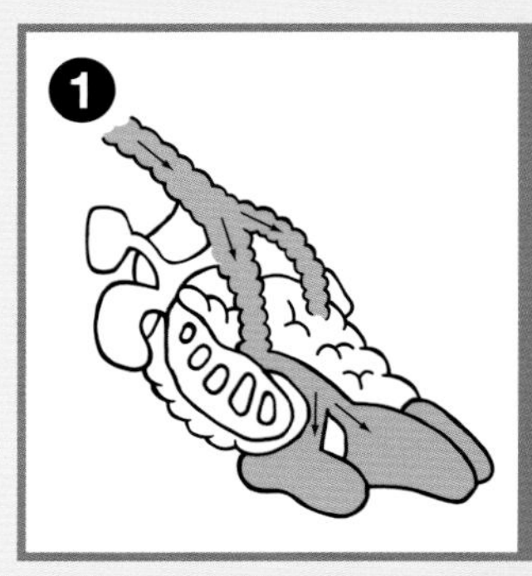

At the first inhalation, six inspiratory muscles move the ribs and sternum outward. This expands the chest cavity, which creates negative pressure inside the bird and causes air to flow into the respiratory tract—through the trachea and mesobronchus. Some of the fresh air goes into the lungs, but most of the air goes into the posterior air sacs (abdominal and caudal thoracic).

At the first exhalation, nine expiratory muscles cause the ribs and sternum to move inward, creating an increase in internal pressure within the air sacs. Air is forced from the posterior air sacs into the lungs. The major bronchi split into thousands of tiny parabronchi that permeate the avian lung tissue.

Parabronchi in the bird's lungs are thinner and smaller than the alveoli in a reptile or mammal's lung, allowing oxygen to pass more easily into the bloodstream. Also, air flows through the parabronchi in one direction, while blood flows through the lung capillaries in the opposite direction, resulting in a remarkably efficient uptake of oxygen.

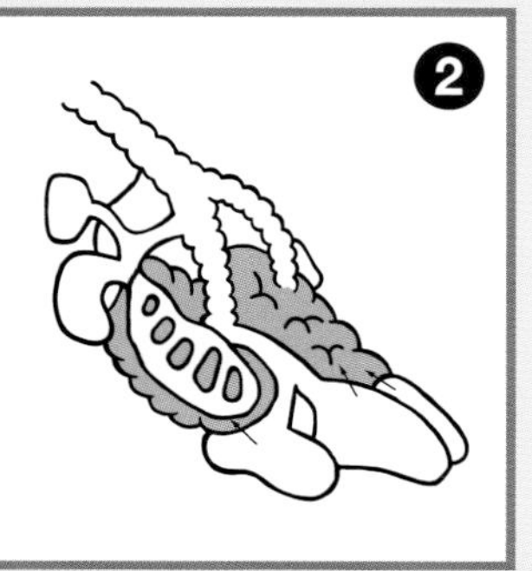

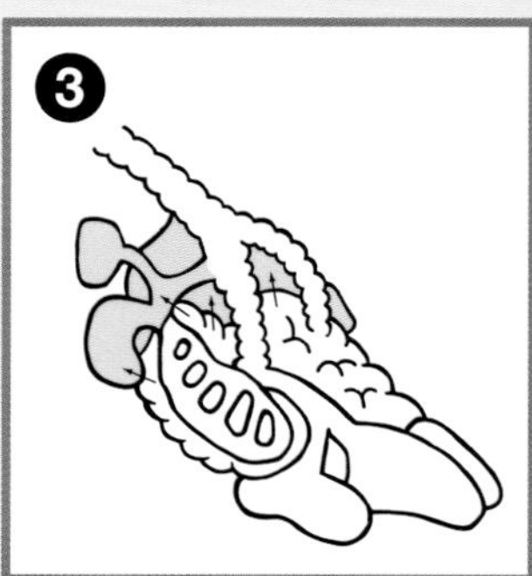

When the bird inhales the second time, negative pressure causes the "used" air to flow from the lung into the anterior air sacs (cranial thoracic, cervical, and clavicular). Along the way, the parabronchi rejoin toward the "exit" end of the lung. Air flows in one direction only, through the lungs.

The second exhalation forces the air out of the anterior air sacs, and out through the trachea.

Not only are air sacs unique to birds, but the structure of the avian lung and the function of the avian respiratory system are quite unique. We know of no breathing system in any other vertebrate species that closely resembles the avian system.

On the other hand, although the avian system is unique among the vertebrates, it is functionally identical in all birds from humming birds to hawks, from robins to ostriches.

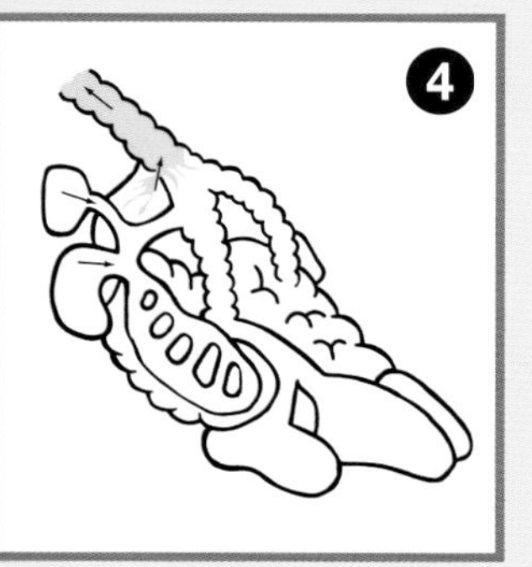

Four Engineering Problems

First, any plausible scenario for the evolution from the reptilian breathing system to the avian breathing system needs to account for the gradual, step-by-step development of a new and unique structure: air sacs. Not just one, but nine air sacs in five variations—abdominal, caudal thoracic, cranial thoracic, cervical, and clavicular. That's hard enough to imagine from a strictly engineering point of view, but there's also the question of timing. Which would arise first, the new air sacs or the four-stage breathing cycle that uses the air sacs? If the air sacs came first, what was their advantageous (selectable) function before four-stage breathing arose? If the air sacs came later, how did four-stage breathing arise without them?

Second, we also need to explain the transition from a two stage to a four stage breathing system. Was there once a three stage breathing system in between? What would that look like? Would there be two inhales for every exhale, or two exhales for every inhale?

Third, the radical transformation of the lung also requires an explanation. What would the intermediate forms between the single opening (in-and-out) reptilian lung and a dual opening (flow through) avian lung look like? How would it happen in small yet advantageous steps? Can there even *be* a transition between a single-opening and a dual-opening system? How would the balloon-like alveoli transform into the tube-like parabronchi? How would the lung maintain function? Would the lung transformation happen before or after the development of air sacs? Would it be before or after the four stage breathing cycle?

Finally, what happens to the diaphragm? The reptiles thought to be the ancestors of birds almost certainly had a diaphragm breathing system.[8] According to many evolutionary biologists, changing from a diaphragm lung system to a flow-through lung would require changing and increasing the musculature of the reptile's chest. At the same time, the diaphragm would need to gradually go away. This poses a

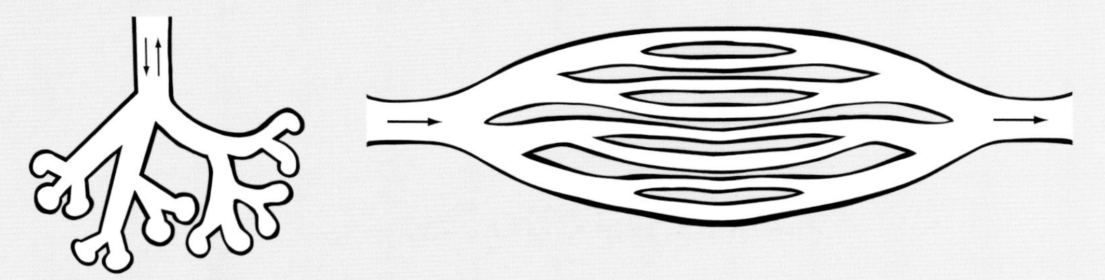

Left The balloon-like alveoli of the reptilian and mammalian lung. Air flows in and out.
Right The tube-like parabronchi of the avian lung. Air flows in and through.

fundamental problem, however. Evolutionary biologist John Ruben points out that the earliest stages of this transformation would have required a hole or hernia in the reptile's diaphragm. This would have immediately compromised the entire system and led to certain death for any animal unfortunate enough to possess this non-functioning intermediate structure. After all, if an animal can't breathe, it will die. And if it dies, it can't reproduce and have offspring that will evolve any further.[9] Can the diaphragm go away gradually? Is a hole in the diaphragm survivable? Let's go back to our do-it-yourself lung and try it.

Do-It-Yourself Lung: Part 2

Carefully make a hole in the thin rubber at the base of the cup. Press on the diaphragm with your thumb, just like you did before.

Watch the balloon as you "exhale" now. Any luck? Now, release the diaphragm and "inhale." Does the balloon inflate like it did before?

Back to the Fossils

Recently paleontologists have found a theropod dinosaur, *Majingatholus atopus*, with features (such as caudal air sacs) that resemble the avian breathing system. If this finding is confirmed, it may well show that some dinosaurs had a breathing system similar to what modern birds have. Some scientists think this finding may help to solve the problem of the origin of the avian breathing system, since many evolutionary biologists think birds evolved from theropod dinosaurs.

Other scientists think that even if this finding is confirmed, all it does is to relocate the problem. At some point, a novel mode of respiration arose. However, the ancestors of the theropod dinosaurs (the reptiles) did not possess this system, so that still leaves us with the questions of how and where such a system arose. It doesn't really matter whether the system first arose in birds or in theropod dinosaurs, it would still need to evolve from a group having a diaphragmatic, two-stage breathing system. The problems of functional transformation we talked about before still need to be solved, whether in a reptile-to-bird transition or in a reptile-to-theropod dinosaur transition. In other words, we still have the problem of maintaining continuous respiratory function during a series of anatomical transformations. As biologist Michael Denton has noted, "Just how such an utterly different respiratory system could have evolved gradually from the standard vertebrate design is fantastically difficult to envisage, especially bearing in mind that the maintenance of respiratory function is absolutely vital to the life of an organism to the extent that the slightest malfunction leads to death within minutes."[10]

So, let's step back a minute. Where do we stand? When scientists compare reptile, avian, and mammal organ systems in the same way they compare skeletal structures, what do they see? Do they see evidence of a gradual transformation of the reptile organ system into an avian organ system? Or into a mammalian system?

Unfortunately, the fossil record is almost entirely mute about these questions. Neither the neo-Darwinists nor their critics have much fossil evidence of the internal organs of any "intermediates."[11] For now, all that we can say for sure is that the internal organs of *living* reptiles, *living* birds, and *living* mammals are very different from one another.

Fossils can tell us many things. From fossil evidence, for example, we know that the first reptiles appear before the first mammals. However, knowing that reptiles came before mammals is not the same thing as knowing that reptiles *produced* mammals. And so the debate continues.*

* For more information, check out **www.exploreevolution.com**

Endnotes

1 David Swift, *Evolution Under the Microscope: A Scientific Critique of the Theory of Evolution,* (Leighton Academic Press: Stirling, Scotland, 2002).

2 Here are some other features that mammals have and reptiles do not: hair or fur, a three-boned inner ear, a single-boned jaw, and highly differentiated teeth.

3 T.S. Kemp, *The Origin & Evolution of Mammals* (New York: Oxford University Press, 2005):129. If such a transition did occur, "…it affected virtually every physiological and anatomical feature."

4 Isaac Salazar-Ciudad, "On the origins of morphological disparity and its diverse developmental basis," *BioEssays* 28 (November 2006):1112-1122.

5 Stephen C. Meyer, "The origin of biological information and the higher taxonomic categories," *Proceedings of the Biological Society of Washington* 117 (2004):213-239.

6 Wallace Arthur, The Origin of Animal Body Plans: A Study in Evolutionary Developmental Biology, (Cambridge: Cambridge University Press, 1997).

see also G.B. Müller and S.A. Newman, "Origination of organismal form: the forgotten cause in evolutionary theory," in G.B. Muller and S.A. Newman, eds. Origination of Organismal Form: Beyond the Gene in Developmental and Evolutionary Biology (Cambridge: The MIT Press, 2003):3-12.

see also C. Nusslein-Volhard and E. Wieschaus, "Mutations affecting segment number and polarity in Drosophila," Nature 287 (1980):795-801.]

7 Michael Denton, *Evolution: A Theory In Crisis* (Burnett Books Limited: London, 1985).

8 "Therapod dinosaurs like modern crocodiles, probably possessed a bellows-like septate lung, and that lung was probably ventilated… by a hepatic-piston diaphragm." John A. Ruben, Terry D. Jones, Nicholas R. Geist, W. Jaap Hillenius, "Lung structure and ventilation in theropod dinosaurs and early birds," *Science* 278 (14 Nov 1997):1268-1269.

9 *"Recently, conventional wisdom has held that birds are direct descendants of theropod dinosaurs. However, the apparently steadfast maintenance of hepatic-piston diaphragmatic lung ventilation in theropods… poses fundamental problems for such a relationship. The earliest stages in the derivation of the avian abdominal air sac system from a diaphragm-ventilating ancestor would have necessitated selection for a diaphragmatic hernia in taxa transitional between theropods and birds. Such a debilitating condition would have immediately compromised the entire pulmonary ventilatory apparatus and seems unlikely to have been of any selective advantage."*

Quoted from John A. Ruben, Terry D. Jones, Nicholas R. Geist, W. Jaap Hillenius, "Lung structure and ventilation in theropod dinosaurs and early birds," *Science* 278 (14 Nov 1997):1269.

10 Michael Denton, *Evolution: A Theory In Crisis* (Burnett Books Limited: London, 1985):211-212.

11 David B. Kitts, "Paleontology and evolutionary theory," *Evolution* 28 (September 1974):458-472.

see also Willem J. Hillenius and John Ruben, "The evolution of endothermy in terrestrial vertebrates: Who? When? Why?" *Physiological and Biochemical Zoology* 77 (2004):1019-1042.]

Conclusion

The Nature of Dissent in Science

THE NATURE OF DISSENT IN SCIENCE

When one hears about a biologist who questions the theory of Universal Common Descent, one might typically assume that he (or she) rejects evolution altogether. Conversely, one might assume that every biologist who accepts Common Descent is also a neo-Darwinist. But the actual diversity of views about the history of life is far more complicated and interesting than that.

Here are a couple of published statements by biologists who have thought very hard about how best to explain the history of life.

Statement A:

"The phenomenon of a monophyletic [single] origin for the universal Tree of Life probably did not occur... At the macro-scale life appears to have had many origins."

Statement B:

"I find the idea of common descent (that all organisms share a common ancestor) fairly convincing, and have no particular reason to doubt it."

Try to guess which scientist is an evolutionary biologist and which scientist is a well known skeptic of neo-Darwinism. Ready for the answer? Statement A was made by Malcolm Gordon, an evolutionary biologist at UCLA. Statement B was made by Michael Behe, a Lehigh University biochemist who is a prominent critic of the power of the mutation/selection mechanism.

If this surprises you, it should. It is easy to lump people together—the jocks, the computer geeks, the motorheads, and the student council types. Unfortunately, when we do this, we may miss the important subtleties and differences that distinguish our individual perspectives on the world. The same is true of how we divide scientists and their views about the history of life. The media often make the situation worse. They, like most people, enjoy a good drama, with well-defined "good guys," "bad guys," and sharply divided opinions that are easy to describe. So we end up with "the creationists versus the evolutionists," a familiar and predictable storyline that, sadly, rolls right over most of the fascinating (and relevant!) details about what individual scientists may actually think. Would you have guessed that an evolutionary biologist would disagree with Universal Common Descent? The conventional media storyline doesn't usually make room for maverick opinions like that—but science itself should.

Another problem arises when dissenting scientists quote the work of their colleagues, many of whom question certain aspects of neo-Darwinism, or parts of the case for it, while still happily calling themselves "evolutionary biologists" or "neo-Darwinists."

Consider the case of a biologist whose research leads her to publish a paper skeptical of the peppered moth story, but who continues to affirm both major tenets of neo-Darwinism. This scientist might resent seeing her skepticism about peppered moths being used as part of a dissenter's case against neo-Darwinism as a whole. Often, in such cases, dissenters are accused of "misquotation" or "misrepresentation." But is this really true?

At the beginning of this book, we told you that historical scientists act like detectives. They observe present-day facts or clues, then try to reconstruct nature's history using what they know about the evidence and about cause-and-effect relationships. We have called many of these "detectives" as witnesses in the current debate.

Witnesses are called to testify because they know something relevant to the case. The duty of an honest witness in court is to state facts or reveal evidence, without regard to whatever his or her opinion may be about the case as a whole. In this book, just as in a court trial, a witness might also be a friend of the defendant, making it difficult to present evidence impartially. Though convinced of the friend's innocence, this witness may still be called upon to reveal incriminating evidence.

So please keep in mind: Just because a scientist is cited or quoted in the "case for" section of one issue, that does not mean that he or she agrees with the whole neo-Darwinian explanation. Conversely, a scientist cited in the "reply" section of this book may still consider him- or herself a neo-Darwinist or an "evolutionist" in some sense of that word. In each case, the scientist is being quoted because he or she has some important evidence to give.

Science is more than simply discovering facts; science is also about interpreting those facts. Part of the scientific process is to debate and argue about which interpretation of the facts best explains what we know. That is why scientists must be free to use all knowledge that has been gathered. Although we give credit to those who make scientific discoveries, in some sense, this knowledge does not "belong" to anyone. Scientists must be free to cite evidence even if their interpretations of the evidence do not agree with those of the original discoverer.

Practicing science should be about making a vigorous effort to make true statements about the natural world, using all the evidence we have gathered, whatever its source, wherever it leads.

By the way...

By now, we hope you can see that real science as it's actually practiced can be a very lively subject. Scientists often disagree about how to interpret the evidence that has come to light, and we hope that this glimpse into how science really works will give you a greater interest in science as a whole. ■

GLOSSARY

adaptation: a feature of an organism that enables it to survive and reproduce in a specific environment.

adenine: one of the four bases in the nucleotides of DNA, commonly denoted by the letter "A."

adulthood: the mature stage of an organism in its life cycle, usually meaning that it is able to reproduce.

allele: alternative form of the same gene locus; one of a pair of genes that occupy corresponding positions on homologous chromosomes and determine alternative expressions of a single trait.

allele frequency: the proportion of a particular allele in a population as a percentage of the total alleles at that locus; also called gene frequency.

allopatric speciation: the process by which, in theory, a new species originates when a population that is geographically separated from the rest of the species becomes unable to reproduce with the original population.

amino acid: an organic compound that contains one or more amino groups and one or more acidic carboxyl groups; amino acids can be combined in chains (polymerized) to form peptides and proteins.

Amphibia: a class of cold-blooded, smooth-skinned vertebrates with legs; offspring usually hatch as aquatic larva with gills, which in most cases metamorphose into adults with air-breathing lungs; includes frogs and salamanders.

analogous structure: a body part in two or more species that performs a similar function but has a different structure; for example, the wings of bats and butterflies.

Animalia: in biological classification, the kingdom comprising all multicellular animals.

angiosperm: a flowering plant that forms its seeds in a protected ovary.

Annelida: a phylum of animals that have segmented bodies with a one-way digestive tract, a circulatory system and a nervous system; includes earthworms.

antibiotic: a chemical produced by one microorganism that can kill or inhibit the growth of other microorganisms; for example, penicillin and streptomycin.

antibiotic resistance: the ability of a microorganism to avoid the harmful effects of an antibiotic by destroying it, transporting it out of the cell, or undergoing changes that block its effects.

Archaea: one of the three domains of life; consists of single-celled organisms without a nucleus (prokaryotes) that differ from Eubacteria in their cell membranes, ribosomes and RNA; includes organisms that thrive in extreme environments such as high salt or heat.

Archaeopteryx: An extinct toothed bird that had a bony, feathered tail and claws.

Arthropoda: a phylum of animals that have a chitinous exoskeleton and a segmented body with paired, jointed appendages; includes insects, crustaceans, arachnids, centipedes and millipedes.

artifact hypothesis: the idea that the Cambrian animal phyla had ancestors, but those ancestors either left no fossil record or have not been found.

artificial selection: the process by which humans deliberately choose to breed only those organisms in a population that have desirable traits.

Aves: a class of warm-blooded feathered vertebrates that have wings for forelimbs; includes both flying and flightless birds.

base pair: two nucleotides on opposite but complementary strands in a DNA molecule; because of their shapes, adenine (A) forms a base pair with thymine (T) and guanine (G) forms a base pair with cytosine (C).

biogeography: the study of the geographical distribution of species.

body plan: the basic symmetry and architecture of an organism; the distinctive anatomical arrangement of fundamental structural elements such as the skeleton or shell; circulatory, respiratory and nervous systems; digestive tract; and appendages.

Burgess Shale: a Middle Cambrian (about 515 million years ago) deposit in British Columbia that contains exquisitely preserved fossils, including many soft-bodied organisms.

bushing: A cylindrical lining used to reduce friction or guide motion.

Cambrian: the first geological period of the Paleozoic Era, lasting from about 540 to 488 million years ago; named after Cambria, Wales.

Cambrian Explosion: an event at the beginning of the Cambrian, lasting less than 10 million years, during which most of the major animal phyla first appear in the fossil record.

Cesarean section: a surgically-assisted birth in which an incision is made in the walls of the mother's abdomen and uterus and the baby is removed without passing through the birth canal; also called C-section.

cassette mutagenesis: a procedure that systematically alters individual DNA codons to determine the effects of those alterations on protein folding or function.

Carnivora: an order of mammals that have large, sharp teeth and powerful jaws and prey on other animals; includes cats, dogs, bears, and weasels.

catalyst: a chemical that increases the rate of a chemical reaction but is not permanently changed by it.

Cephalopod: a class of marine mollusks that have tentacles, horny jaws and well-developed eyes and nervous systems; includes octopuses, squids, and *Nautilus*.

Cetacea: an order of marine mammals; includes whales, dolphins and porpoises.

chance: the characteristic of lacking any discernible pattern or direction, with more than one outcome being possible in a given set of circumstances.

cell: the basic structural and functional unit of all living organisms, enclosed by a semipermeable plasma membrane.

Chengjiang biota (or fauna): a suite of fossils from the lower Cambrian Maotianshan shale in China, characterized by exquisite soft-body preservation.

Chiroptera: an order of flying mammals; includes bats.

Chordata: a phylum of animals that have a dorsal nerve cord and a notochord during development; includes the vertebrates (animals with backbones).

chromosome: a thread-like structure in the nucleus of a eukaryotic cell; consists of a single compacted molecule of DNA and some proteins.

cilia: short, cylindrical projections on living cells that typically function in locomotion.

circular definition: a logical fallacy in which the term being defined is used as part of the definition.

cladogram: a branching diagram that arranges organisms according to their shared and unshared characteristics; often used to construct hypotheses about the ancestor-descendant relationships of those organisms.

class: the level of biological classification above an order and below a phylum.

codon: The basic unit of the genetic code; a sequence of three adjacent nucleotides in DNA or mRNA that specifies one amino acid.

collagen: a long fibrous protein that is the main constituent of connective tissues and the most abundant protein in animals.

compensatory mutation: a secondary mutation in a bacterium that helps to restore the original function lost due to a prior mutation that rendered the bacterium resistant to an antibiotic.

competitive advantage: the increased ability of an organism to survive and reproduce in comparison with other organisms competing for limited resources.

complexity: the improbability of assembling a structure, system, or molecule.

contingency: characteristic of an event that is only one of several logically or physically possible outcomes.

convergent evolution: the appearance of similar characteristics in two or more taxa by independent lines of descent.

co-option: the use of an existing biological structure or feature for a new function; also called exaptation.

Cretaceous: a geological period lasting from about 145 to 65 million years ago, at the end of which dinosaurs became extinct.

cytochrome c: an iron-bearing molecule utilized in the electron transport chains of mitochondria and chloroplasts.

cytosine: one of the four bases in the nucleotides of DNA, commonly denoted by the letter "C."

cytoskeleton: a network of microscopic fibers that stabilize the shape of a eukaryotic cell and function in intracellular transport.

cytoplasm: all the contents of a cell inside the plasma membrane (but, in eukaryotes, outside the nucleus).

Darwinism: the theory that all living things are descended from a common ancestor and modified by unguided natural processes such as natural selection and random variation.

developmental pathway: the route or process by which a structure or organ forms in the embryo of an organism.

diaphragm: a large, thin sheet of muscle that seals the airtight sac or chamber containing the lungs, and completely separates the chest cavity from the abdomen.

differential reproduction: variation in the rate of producing offspring between organisms that are better adapted to their environment and those that are less well adapted.

disparity: major variations in morphology or body plan.

diversity: minor variations within a basic body plan or biological form; for example, variations among dogs such as a golden retriever and St. Bernard.

DNA: deoxyribonucleic acid; a molecule consisting of two long, intertwined chains of nucleotides that carry the information to specify the sequence of amino acids in proteins.

domains: taxonomic categories higher than kingdom, defined by the gene sequences and basic mechanisms used to perform fundamental cellular functions; the three domains are the Archaea, Eubacteria, and Eukaryota.

duplication mutation: the production of a copy of some segment of DNA during meiosis.

Echinodermata: a phylum of marine animals that are radially symmetrical as adults and have a calcareous endoskeleton, tube feet and water vascular system; includes starfishes and sea urchins.

Ediacaran fauna: a group of Precambrian (Vendian) multicellular organisms, named after the Ediacaran Hills in Australia.

embryo: an early developmental stage of a multicellular organism.

enzyme: a protein that functions as a catalyst in biochemical reactions.

Eocene: a geological period lasting from about 56 to 34 million years ago, marked by the diversification of mammals.

Eubacteria: one of the three domains of life; consists of single-celled organisms without a nucleus (prokaryotes) that differ from Archaea in their cell membranes, ribosomes and RNA; includes gram-negative bacteria such as E. coli and gram-positive bacteria such as Staphylococcus and Streptococcus.

eukaryote: a cell containing membrane-bound organelles, including (most importantly) a nucleus.

Eukaryota (also known as Eukarya): one of the three domains of life; includes all organisms composed of one or more cells that contain a distinct membrane-bound nucleus.

evolution: of the many meanings of this word, three are used here: (1) *change over time;* the fact that most of the organisms alive today are different from organisms that existed in the past; (2) *universal common descent;* the hypothesis that all organisms are modified descendants of a single common ancestor in the distant past; (3) *the mechanisms of biological change;* the hypothesis that natural selection acting on random variations has been the principal cause of modification.

exaptation: the use of an existing biological structure or feature for a new function; co-option.

exoskeleton: a stiff external covering that contains, supports, and protects an animal's body.

extrapolation: a conjecture based on the assumption that a phenomenon or trend observed in the present can be extended into the past or future.

family: the level of biological classification above a genus and below an order.

fitness: the ability of an organism to survive and produce viable offspring in a given environment.

fitness cost: the decrease in an organism's ability to survive and reproduce in other environments following a mutation that confers a selective advantage in one environment.

flagellum: a long whiplike appendage that provides some function (usually locomotion) in microorganisms.

fossil: the mineralized remains, impression, or trace of a once-living organism.

fossil succession: the specific order of fossils, from lower to higher, within geological strata.

function: the role that a biological structure or feature plays in survival, reproduction or other activities of an organism.

functional information: nucleotide sequences of DNA and RNA that code for proteins with biological functions.

Galápagos Archipelago: a group of volcanic islands about 600 miles off the west coast of Ecuador, South America.

gene: a unit of heredity on a chromosome, usually understood as a DNA sequence that specifies a particular protein.

gene pool: the total genetic material in a population at a given time.

genetic information: the sequences of nucleotides in DNA and RNA that specify the sequences of amino acids in proteins.

genome: the total DNA of an organism.

genotype: the combination of alleles inherited for a particular trait.

genus: the level of biological classification above a species and below a family.

geographic barrier: any physical feature of the earth that separates two populations and prevents interbreeding between them.

Gondwanaland: A prehistoric supercontinent in the southern hemisphere that was broken up by the action of plate tectonics to form the modern continents.

gradualism: the Darwinian view that since all species have descended from other species by the ordinary process of reproduction, evolution has occured in steps no larger than those that now distinguish parents and offspring.

guanine: one of the four bases in the nucleotides of DNA, commonly denoted by the letter "G."

halteres: tiny structures behind the wings in some insects (such as fruit flies), which help to stabilize the animal in flight; also called balancers.

hemoglobin: an iron-containing protein that carries oxygen in the blood.

heredity: the transmission of traits from parents to offspring.

heritability: the proportion of phenotypic variation in a population that is attributable to differences in genotype.

hexokinase: an enzyme that catalyzes the addition of a phosphate group ("phosphorylation") to a six-carbon sugar ("hexose"), a vital step in metabolism.

historical science: an enterprise that observes and studies clues left by past events and uses what is known about present cause-and-effect relationships to reconstruct the history of those events; examples include geology, paleontology, archaeology and forensics.

homeotic gene: a DNA sequence that affects embryo development by specifying the character of a body segment; for example, *Antennapedia,* in which a mutated gene can cause a fly to grow a leg from its head instead of an antenna.

homologous structure: a body part that is similar in structure and position in two or more species but has a different function in each; for example, the forelimbs of bats, porpoises and humans.

homoplasy: structural similarity not thought to be due to inheritance from a common ancestor; see convergent evolution.

hopeful monster: a hypothetical organism that supposedly originates in a single generation from a "macromutation" with large-scale effects on anatomy.

Hox gene: one of a cluster of homeotic genes; the Hox genes of many animals (such as insects and mammals) are remarkably similar.

humerus: in humans, a long bone in the upper arm that extends from the shoulder to the elbow; also, the corresponding bone in other animals.

Hymenoptera: an order of insects that includes bees, wasps and ants.

hypothesis: (1) an educated guess; (2) a tentative explanation; (3) a proposition to be tested by comparing it to evidence.

inference to the best explanation: a method of scientific reasoning that favors the hypothesis that would, if true, best explain the relevant evidence; hypotheses that qualify as "best" typically provide coherent and causally adequate explanations of the evidence or phenomenon in question.

Insecta: a class of arthropods characterized by a body divided into three parts (head, thorax, and abdomen); the head has compound eyes and the thorax has three pairs of legs.

irreducible complexity: the characteristic of a system of well-matched, mutually interacting parts performing a specific function, in which the removal of any one of the parts causes the system to cease functioning.

kingdom: the highest level of biological classification below a domain; the domain Eukaryota includes the kingdoms Protista, Fungi, Plantae and Animalia.

LUCA: Last Universal Common Ancestor; the most recent organism from which, in theory, all subsequent organisms have descended.

macroevolution: the origin of new large-scale features such as organs or body plans.

maladapted: poorly suited to survive and reproduce in a given environment.

Mammalia: A class of warm-blooded, fur-bearing vertebrates characterized by mammary glands, with which females produce milk to feed the young.

mammal-like reptiles: an extinct group of reptiles that first appeared during the Permian period.

manus: the most distal part of the forelimb of a vertebrate; in humans, this includes the wrist and hand.

marsupial: an animal (usually a mammal) in which embryos complete their development in a maternal pouch called a "marsupium."

mechanism: the process by which something occurs.

microevolution: small-scale changes within existing species.

minimal complexity: the simplest state, either genetic or metabolic, that is consistent with the viability of an organism.

mitochondria: organelles in eukaryotic cells that convert food into a form of energy usable by the cell.

molecular homology: similarity of the nucleotide sequences of DNA or RNA molecules, or the amino acid sequences of proteins.

molecular machines: microscopic assemblages of biological molecules (such as proteins) that move and perform key cellular functions.

Mollusca: a phylum of animals that are unsegmented and soft-bodied, and have a muscular foot and often one or more hard shells; includes clams, snails and octopuses.

monophyletic: descended from a single common ancestor.

morphology: the form or structure of an organism; anatomy.

mRNA: messenger RNA; the sequence of mRNA contains the same information as the DNA sequence in a "gene" and specifies the order of amino acids in a protein.

mutant: an organism with a new trait resulting from a mutation.

mutation: alteration of an organism's DNA due to mistakes during replication or damage from external agents such as chemicals or radiation.

natural selection: the process in which organisms better adapted to their environment survive and reproduce at a higher rate than those less adapted, with the result that the survivors' characteristics are more prevalent in subsequent generations.

Nautilus: a cephalopod with an external spiral shell.

necessity: characteristic of an event in which there were no other logically or physically possible outcomes.

Nematoda: a phylum of animals that have cylindrical, bilaterally symmetrical bodies; includes roundworms and hookworms.

neo-Darwinism: the modern version of Darwinian evolutionary theory, according to which new variations originate in DNA mutations that provide the raw materials upon which natural selection may act to produce evolutionary change.

notochord: a flexible column located between the gut and nerve cord in the embryos of all chordates, a group of animals that includes the vertebrates (animals with backbones).

nucleic acid: a molecule consisting of joined nucleotides that can carry biological information; for example, DNA and RNA.

nucleotide: the fundamental structural unit of a nucleic acid (DNA or RNA); consists of a nitrogen-carrying base (purine or pyrimidine), a sugar molecule, and a phosphate group.

ontogeny: the development of an organism over its lifetime, from conception to death.

order: the level of biological classification above a family and below a class.

ORFan genes: DNA sequences that resemble no other known genes in biology and are thus evolutionarily untraceable.

organ: a group of tissues that form a distinct structure and perform a specific function, such as a heart or lung.

paleontology: the scientific study of fossils and their use to reconstruct the history of life.

parabronchi: tiny air vesicles in a bird's lung where gas exchange occurs.

penicillin: an antibiotic naturally produced by some molds of the genus *Penicillium* that interferes with the construction of new bacterial cell walls.

penicillinase: a complex enzyme found in some bacteria, which destroys penicillin and confers resistance to it.

peppered moths: *Biston betularia,* a species of night-flying moth sometimes cited in textbooks as an example of evolutionary change produced by natural selection.

pharyngula: a developmental stage in vertebrate embryos, after fertilization, cleavage and gastrulation, in which the embryos are characterized by a notochord, a post-anal tail, and a series of paired folds in the neck region.

phenotype: the observable physical characteristics of an organism or the physical manifestation of some trait, such as blood type, protein structure or body plan.

phylogeny: the evolutionary history of a group of organisms.

phylum: the highest level of biological classification within a kingdom.

point mutation: the substitution of one nucleotide for another in DNA or RNA.

polypeptide: a molecule consisting of many amino acids joined by chemical bonds between their amino and carboxyl groups; not necessarily biologically active.

polyphyletic: descended from different ancestors.

population: a group of individuals of one species in the same geographical area.

prebiotic: before the existence of biological life.

Precambrian: the geological time in Earth's history before the beginning of the Cambrian period (about 540 million years ago).

primates: an order of mammals with binocular vision, specialized appendages for grasping and enlarged brains; includes lemurs, monkeys, apes and humans.

protein: a large polypeptide that performs a biological function.

protoplasm: the substance inside a living cell, once thought to be simple, but now known to be a complex network of biomolecules, microscopic structures and molecular machines; includes the cytoplasm and (in eukaryotes) the nucleus.

punctuated equilibrium: a characteristic of the fossil record in which new species appear suddenly (punctuation), then persist unchanged (stasis) until they disappear (extinction); named by Stephen Jay Gould and Niles Eldredge and attributed by them to allopatric speciation.

random variation: minor differences among the individual organisms in a population.

radius: in humans, a bone in the forearm that extends from the inside of the elbow to the thumb side of the wrist; also, the corresponding bone in other animals.

respiratory system: the organs in animals used for gas exchange.

reproductive success: the ability of an organism to reproduce and pass on its genes to offspring.

ribosome: an organelle composed of protein and RNA that uses the information in mRNAs to synthesize proteins inside a cell.

RNA: ribonucleic acid; a molecule consisting of the nucleotides adenine, guanine, cytosine and uracil (instead of thymine); involved in converting DNA sequences into amino acid sequences during protein synthesis in living cells.

scute: a protective horny plate on the exterior of a reptile.

selective advantage: the characteristic of an organism that enables it to survive and reproduce better than other organisms in a population in a given environment.

selective predation: a behavior in which a predator attacks only certain types of individuals in a population of potential prey.

speciation: the origin of a new species from an existing one.

species: the taxonomic rank below genus; there are many definitions of this word, all of which are controversial to some extent, but the most common definition used for sexually reproducing organisms is "a group of interbreeding organisms that is reproductively isolated from other such groups."

stasis: the persistence of a particular species without discernible change through geological strata.

stator: a stationary part of a machine that remains fixed while other parts rotate around it.

sternum: the breastbone in vertebrates; a long, flat bone located in the middle of the chest.

strata: layers of rock, typically horizontal.

sympatric speciation: the process by which, in theory, a new species originates from another without being geographically separated from it.

taxon: a group or category of biological classification at any level (plural: taxa).

tetrapod: a vertebrate animal with four legs or leg-like appendages.

theropod: a group of terrestrial, bipedal dinosaurs that were primarily carnivorous; one example was *Tyrannosaurus rex.*

thymine: one of the four bases in the nucleotides of DNA, commonly denoted by the letter "T."

tissue: a group of interconnected cells forming a structure or performing a particular function in a multicellular organism.

transcription: the process by which mRNA is synthesized using DNA as a template.

transitional form: an organism that is intermediate between an ancestor and a descendant in a evolutionary sequence.

translation: the process by which protein is synthesized by a ribosome using mRNA as a template.

trait: any characteristic of an organism, whether genetic or structural.

Tree of Life: Darwin's metaphor for the history of life, which portrays all living things (the tips of the branches) as modified descendants of a single common ancestor (the root or trunk).

trunk: the body of a vertebrate animal excluding the head, limbs and tail.

type III secretory system: a microscopic needle-like structure that some bacteria use to inject toxins into other cells; abbreviated TTSS.

ulna: in humans, a bone in the forearm that extends from the outside of the elbow to the side of the wrist opposite the thumb; also, the corresponding bone in other animals.

universal common ancestry: the hypothesis that all species are biological descendants of a single common ancestor, and that all life can be portrayed as a tree with a single trunk or root.

variation: one or more differences among individual organisms in the same species.

Vendian: the geological period just before the Cambrian, lasting from about 650 to 540 million years ago; contains fossils of the Ediacaran fauna.

vertebrates: chordates with a backbone; these include fish, amphibians, reptiles, birds, and mammals.

viability: the ability of an organism to survive.

Selected Bibliography

Aboitiz, F., (1988), Homology: a Comparative or a Historical Concept? *Acta Biotheoretica* 37: 27–29.

Ager, D. V., (1976), The Nature of the Fossil Record, *Proceedings of the Geological Association* 87: 131–159.

Alberts, B., (1998), The Cell as a Collection of Protein Machines: Preparing the Next Generation of Molecular Biologists, *Cell,* February 8, 1998, 92: 291–294.

Andersson, D. I. & Hughes, D., (1996), Muller's Ratchet Decreases Fitness of a DNA-based Microbe, *Proceedings of the National Academy of Sciences,* January 23, 1996, 93: 906–907.

Aris-Brosou, S. & Yang, Z., (2003), Bayesian Models of Episodic Evolution Support a Late Precambrian Explosive Diversification of the Metazoa, *Molecular Biology and Evolution* 20: 1947–1954.

Arthur, W., (2004a), *Biased Embryos and Evolution* (Cambridge, England: Cambridge University Press).

Arthur, W., (2004b), The Effect of Development on the Direction of Evolution: Toward a Twenty-first Century Consensus, *Evolution & Development* 6: 282–288.

Arthur, W., (1997), *The Origin of Animal Body Plans: A Study in Evolutionary Developmental Biology* (Cambridge, England: Cambridge University Press).

Axe, D., (2004), Estimating the Prevalence of Protein Sequences Adopting Functional Enzyme Folds, *Journal of Molecular Biology* 20: 1–21.

Axelsson, M., (2001), The Crocodilian Heart: More Controlled than We Thought? *Experimental Physiology* 86: 785–789.

Ayala, F., (1994), Darwin's Revolution, in J. H. Campbell & J. W. Schopf (eds.), *Creative Evolution!?* (Sundberg, MA: Jones and Bartlett), 4–5.

Baguñà, J. & Garcia-Fernàndez, J., (2003), Evo-devo: The Long and Winding Road, *International Journal of Developmental Biology* 47: 705–713.

Beck, C. B., (1976), *Origin and Early Evolution of Angiosperms* (New York: Columbia University Press).

Behe, M. J., (1996), *Darwin's Black Box: the Biochemical Challenge to Evolution* (New York: The Free Press).

Behe, M. J., Snoke, D. W., (2004), Simulating Evolution by Gene Duplication of Protein Features that Require Multiple Amino Acid Residues, *Protein Science* 13: 2651–2664.

Berg, L. S., (1969), *Nomogenesis OR Evolution Determined By Law* (Cambridge, MA: The MIT Press).

Bininda-Emonds, O. R. P., Jeffery, J. E., & Richardson, M. K., (2003), Inverting the Hourglass: Quantitative Evidence Against the Phylotypic Stage in Vertebrate Development, *Proceedings of the Royal Society of London,* Biological Science, 270 (1513): 341–346.

Bird, R. J., (2003) *Chaos and Life: Complexity and Order in Evolution and Thought* (New York: Columbia University Press).

Bowring, S. A., Grotzinger, J. P., Isachsen, C. E., Knoll, A. H., Pelechaty, S. M., & Lolosov, P., (1993), Calibrating Rates of Early Cambrian Evolution, *Science* 261: 1293–1298.

Bowring, S. A., Grotzinger, J. P., Isachsen, C. E., Knoll, A. H., Pelechaty, S. M., & Lolosov, P., (1998), A New Look at Evolutionary Rates in Deep Time: Uniting Paleontology and High-Precision Geochronology, *GSA Today* 8: 1–8.

Bradshaw, A. D., (1991), Genostasis and the Limits to Evolution, *Philosophical Transactions of the Royal Society of London,* Biological Sciences, 333 (1267): 289–305.

Brady, R. H., (1979), Natural Selection and the Criteria by which a Theory Is Judged, *Systematic Zoology* 28: 600–621.

Brocchieri, L., (2001), Phylogenetic Inferences from Molecular Sequences: Review and Critique, *Theoretical Population Biology* 59: 27–40.

Campbell, J. A. & Meyer, S. C., (eds.), (2003), *Darwinism, Design, and Public Education* (East Lansing: Michigan State University Press).

Campbell, K. S. W. & Marshall, C. R., (1987), Rates of Evolution among Paleozoic Echinoderms, in K. S. W. Campbell and M. F. Day, (eds.), *Rates of Evolution* (London: Allen and Unwin).

Cao, Y., Waddell, P. J., Okada, N., & Hasegawa, M., (1998), The Complete Mitochondrial DNA sequence of the Shark Mustelus manazo: Evaluating Rooting Contradictions to Living Bony Vertebrates, *Molecular Biology and Evolution* 15: 1637–1646.

Carroll, R. L., (2000), Towards a New Evolutionary Synthesis, *Trends in Ecology and Evolution,* January 2000, 5: 27–32.

Carroll, R. L., (1997–1998), Limits to Knowledge of the Fossil Record, *Zoology* 100: 221–231.

Chien, P., Chen, J. Y, Li, C. W. & Leung, F. (2001). Berkeley, California. North American Paleontological Convention, June 26–July 1. *Observation of Precambrian Sponge Embryos from Southern China, Revealing Ultrastructures Including Yolk Granules, Secretion Granules, Cytoskeleton and Nuclei.* Conf. Pres. Unpl.

Cleland, C., (2002), Methodological and Epistemic Differences Between Historical Science and Experimental Science, *Philosophy of Science* 69: 474–496.

Cleland, C., (2001), Historical Science, Experimental Science, and the Scientific Method, *Geology* 29: 987–990.

Cohen, P., (2003), Renegade Code, *New Scientist* 179: 34–38.

Côté, I. M. & Sutherland, W. J., (1997), The Effectiveness of Removing Predators to Protect Bird Populations, *Conservation Biology,* April 1997, 11: 395.

Coyne, J., (1996), God in the Details: The Biochemical Challenge to Evolution, *Nature*, September 19, 1996, 383: 227–228.

Cracraft, J., (1979), Phylogenetic Analysis, Evolutionary Models, and Paleontology, in Joel Cracraft and Niles Eldredge (eds.), *Phylogenetic Analysis and Paleontology* (New

York: Columbia University Press), 7–39.

Darwin, C., (1964), *On the Origin of Species* (Cambridge, MA: Harvard University Press) Facsimile of the First Edition, 1859.

Davidson, E. H., (2001), *Genomic Regulatory Systems: Development and Evolution* (San Diego: Academic Press).

De Bodt, S., Maere, S. & Van de Peer, Y., (2005), Genome Duplication and the Origin of Angiosperms, *Trends in Ecology and Evolution* 20: 591–597.

Delarbre, C., Rasmussen, A-S., Arnason, U., & Gachelin, G., (2001), The Complete Mitochondrial Genome of the Hagfish Myxine glutinosa: Unique Features of the Control Region, *Journal of Molecular Evolution* 53: 634–641.

Dembski, W. A., (2002) *No Free Lunch* (Lanham, MD: Rowman & Littlefield Publishers).

Dembski, W. A. & Ruse, M. (eds.), (2004), *Debating Design* (Cambridge, England: Cambridge University Press).

Denton, M., (1985) *Evolution: A Theory In Crisis* (London: Burnett Books Ltd.).

de Pouplana, L. R., (ed.), (2004), *The Genetic Code and the Origin of Life* (New York: Kluwer Academic and Plenum Publishers).

Dobzhansky, T., Ayala, F., Stebbins G., & Valentine, J., (1977), *Evolution* (San Francisco: W. H. Freeman).

Doolittle, R. F., (2002), Microbial Genomes Multiply, *Nature* April 18, 2002, 416: 698.

Doolittle, W. F., (2000), The Nature of the Universal Ancestor and the Evolution of the Proteome, *Current Opinion in Structural Biology* 10: 355–358.

Dover, G., (2000), *Dear Mr. Darwin: Letters on the Evolution of Life and Human Nature* (Berkeley: University of California Press).

Droser, M. L., Jensen, S. & Gehling, J. G., (2002), Trace Fossils and Substrates of the Terminal Proterozoic-Cambrian Transition: Implications for the Record of Early Bilaterians and Sediment Mixing, *Proceedings of the National Academy of Sciences* 99: 12572–12576.

Duvall, M. R. & Ervin, A. B., (2004), 18S Gene Trees are Positively Misleading for Monocot/Dicot Phylogenetics, *Molecular Phylogenetics and Evolution* 30: 97–106.

Eldredge, N., (1985), *Time Frames: The Rethinking of Darwinian Evolution and the Theory of Punctuated Equilibria (*New York: Simon & Schuster).

Eldredge, N. & Gould, S. J., (1972), Punctuated Equilibria: An Alternative To Phyletic Gradualism, in T. J. M. Schopf (ed.), *Models In Paleobiology* (San Francisco: Freeman, Cooper and Co.), 82–115.

Endow, S. A., (2003), Kinesin Motors as Molecular Machines, *BioEssays* 25: 1212–1219.

Erwin, D. H., (2000), Macroevolution is More than Repeated Rounds of Microevolution, *Evolution and Development* 2: 78–84.

Erwin, D. H., (1999), The Origin of Bodyplans, *American Zoologist* 39: 617–629.

Fenske, C., Palm, G. J. & Hinrichs, W., (2003), How Unique Is the Genetic Code? *Angewandte Chemie International Edition* 42: 606–610.

Fischer, D. & Eisenberg, D., (1999), Finding Families for Genomic ORFans, *Bioinformatics* 15: 759–762.

Foote, M., (1997), Sampling, Taxonomic Description, and Our Evolving Knowledge of Morphological Diversity, *Paleobiology,* Spring, 1997, 23: 181–206.

Franklin, C., Seebacher, F., Grigg, G. C. & Axelsson, M., (2000), At the Crocodilian Heart of the Matter, *Science*, September 8, 2000, 289: 1687–1688.

Franklin, C. E. & Axelsson, M., (1994), The Intrinsic Properties of an *In Situ* Perfused Crocodile Heart, *Journal of Experimental Biology* 186: 269–288.

Gaffney, B. & Cunningham, E. P., (1988), Estimation of Genetic Trend in Racing Performance of Thoroughbred Horses, *Nature*, April 21, 1988, 332: 722–724.

Gee, H., (1999), *In Search of Deep Time: Beyond the Fossil Record to a New History of Life* (New York: The Free Press).

Gilbert, S. F., Opitz, J. M. & Raff, R. A., (1996), Resynthesizing Evolutionary and Developmental Biology, *Developmental Biology* 173: 357–372.

Gingerich, P., (1994), The Whales of Tethys, *Natural History*, April 1994, 103: 86–88.

Gingerich, P., Raza, S. M., Arif, M., Anwar, M. & Zhou, X., (1994), New Whale From the Eocene of Pakistan and the Origin of Cetacean Swimming, *Nature* 368: 844–847.

Gingerich, P., Smith, B. H. & Simons, E. L., (1990), Hind Limbs of Eocene Basilosaurus: Evidence of Feet in Whales, *Science* 249: 154–157.

Goodwin, B., (1994), *How the Leopard Changed Its Spots: The Evolution of Complexity* (New York: Charles Scribner's Sons).

Goodwin, B., (1986), Is Biology An Historical Science? in S. Rose & L. Appignanesi, eds., *Science and Beyond*, London, Basil Blackwell, 57.

Gordon, M. S., (1999), The Concept of Monophyly: A Speculative Essay, *Biology and Philosophy* 14: 331–348.

Gordon, M. S. & Olson, E. C., (1995), *Invasions of the Land: The Transitions of Organisms from Aquatic to Terrestrial Life* (New York: Columbia University Press).

Gould, S. J., (2000), Abscheulich! Atrocious! *Natural History*, March, 2000, 42–49.

Gould, S. J., (1982), Darwinism and the Expansion of Evolutionary Theory, *Science* 216: 380–387.

Gould, S. J., (1986), Evolution and the Triumph of Homology: Or, Why History Matters, *American Scientist* 74: 61.

Gould, S. J., (1982), Punctuated Equilibrium—A Different Way of Seeing, *New Scientist,* April 15, 1982, 135–141.

Gould, S. J. & Eldredge, N., (1997), Punctuated Equilibria: The Tempo and Mode of Evolution Reconsidered, *Paleobiology* 3: 115–151.

Gould, S. J. & Eldredge, N., (1993), Punctuated Equilibrium Comes of Age, *Nature*, November 1993, 366: 223.

Grant, P. R. & Grant, B. R., (2002), Unpredictable Evolution in a 30-Year Study of Darwin's Finches, *Science* 296: 707–711.

Graur, D. & Martin, W., (2004), Reading the Entrails of Chickens: Molecular Timescales of Evolution and the Illusion of Precision, *Trends in Genetics*, February 2004, 20: 80–86.

Hagadorn, J. W. *et. al.,* (2006) Cellular and Subcellular Structure of Neoproterozoic Animal Embryos, *Science*, October 2006, 314: 291–294.

Hall, B. K., (1996), Baupläne, Phylotypic Stages, and Constraint: Why There Are So Few Types of Animals, *Evolutionary Biology* 29: 215–253.

Hall, B. K., (1995), Homology and Embryonic Development, *Evolutionary Biology* 28: 1–37.

Harold, F. M., (2001), *The Way of the Cell: Molecules, Organisms, and the Order of Life* (New York: Oxford University Press).

Hedges, S. B. & Sibley, C. G., (1994), Molecules vs. Morphology in Avian Evolution: The Case of the 'Pelecaniform' Birds, *Proceedings of the National Academy of Sciences USA*, October 1994, 91: 9861–9865.

Hillenius, W. J. & Ruben, J., (2004), The Evolution of Endothermy in Terrestrial Vertebrates: Who? When? Why? *Physiological and Biochemical Zoology* 77: 1019–1042.

Hooper, J., (2002), *Of Moths and Men: An Evolutionary Tale* (New York: W. W. Norton).

Jablonski, D., (2000), Micro- and Macroevolution: Scale and Hierarchy in Evolutionary Biology and Paleobiology, *Paleobiology* 26: 15–52.

Kemp, T. S., (1982), *Mammal-Like Reptiles and the Origin of Mammals* (London, England: Academic Press).

Kemp, T. S., (2005), *The Origin & Evolution of Mammals* (New York: Oxford University Press).

Kettlewell, H. B. D., (1955), Selection Experiments on Industrial Melanism in the Lepidopotera, *Heredity* 9: 323–342

Kirschner, M. & Gerhart, J., (2005), *The Plausibility of Life: Resolving Darwin's Dilemma* (New Haven: Yale University Press).

Kitts, D. B., (1974), Paleontology and Evolutionary Theory, *Evolution*, September 1974, 28: 458–472.

Kutilek, V. D., Sheeter, D. A., Elder, J. H. & Torbett, B. E., (2003), Is Resistance Futile? *Current Drug Targets—Infectious Disorders*, December 2003, 4: 295–309.

Kutschera, U. & Niklas, K. J., (2004), The Modern Theory of Biological Evolution: An Expanded Synthesis, *Naturwissenschaften* 91: 255–276.

Lee, M. S. Y., (1999a), Molecular Phylogenies Become Functional, *Trends in Ecology and Evolution* 14: 177–178.

Lee, M. S. Y., (1999b), Molecular Clock Calibrations and Metazoan Divergence Dates, *Journal of Molecular Evolution* 49: 385–391.

Lerner, I. M., (1954), *Genetic Homeostasis* (Edinburgh: Oliver and Boyd).

Levinton, J., (1998), *Genetics, Paleontology, and Macroevolution* (Cambridge, England: Cambridge University Press).

Lewis, E. B., (1985), Regulation of the Genes of the Bithorax Complex in Drosophila, *Cold Spring Harbor Symposia on Quantitative Biology* 50: 155–164.

Lewis, E. B., (1978), A Gene Complex Controlling Segmentation in Drosophila, *Nature* 276: 565–570.

Linton, A., (2001), Scant Search for the Maker, *Times Higher Education Supplement*, April 20, 2001, 29.

Long, J. A. & Gordon, M. S., (2004), The Greatest Step in Vertebrate History: A Paleobiological Review of the Fish-Tetrapod Transition, *Physiological and Biochemical Zoology* 77: 700–719.

Løvtrup, S., (1979), Semantics, Logic, and Vulgate Neo-Darwinism, *Evolutionary Theory* 4: 157–172.

Lynch, M., (1999), The Age and Relationships of the Major Animal Phyla, *Evolution* 53: 319–325.

McDonald, J. F., (1983), The Molecular Basis of Adaptation: A Critical Review of Relevant Ideas and Observations, *Annual Review of Ecology and Systematics* 14: 77–102.

McShea, D. W., (1993), Arguments, Tests, and the Burgess Shale—A Commentary on the Debate, *Paleobiology* 19: 399–402.

Medawar, P., (1983) Nobel Conference XVIII, *Darwin's Legacy* (San Francisco: Harper & Row).

Meyer, S. C., (2004), The Origin of Biological Information and the Higher Taxonomic Categories, *Proceedings of the Biological Society of Washington* 117: 213–239.

Meyer, S. C., Ross, M., Nelson, P. & Chien P., (2003), The Cambrian Explosion: Biology's Big Bang, in J. A. Campbell & S. C. Meyer (eds.), *Darwinism, Design, and Public Education* (East Lansing: Michigan State University Press), 323–402.

Miklos, G. L. G., (1993), Emergence of Organizational Complexities During Metazoan Evolution: Perspectives from Molecular Biology, Palaeontology and Neo-Darwinism, *Memoirs of the Association of Australasian Palaeontologists* 15: 7–41.

Miklos, G. L. G & John, B., (1987), From Genome to Phenotype, in K. S. W. Campbell & M. F. Day (eds.), *Rates of Evolution* (London: Allen and Unwin).

Milius, S., (2000), Toothy Valves Control Crocodile Hearts, *Science News*, August 26, 2000, 158: 133.

Mindell, D. P. & Meyer, A., (2001), Homology Evolving, *Trends in Ecology and Evolution* 16: 343–440.

Minnich, S. A. & Meyer, S. C., (2004), Genetic Analysis of Coordinate Flagellar and Type III Regulatory Circuits in Pathogenic Bacteria, in M. W. Collins & C. A. Brebbia (eds.), *Design and Nature II: Comparing Design in Nature with Science and Engineering* (Southampton: Wessex Institute of Technology Press), 295–304.

Moss, L., (2004), *What Genes Can't Do* (Cambridge, MA: The MIT Press).

Müller, G., (1999), Discussion of Wake Paper, in G. R. Bock and G. Cardew, (eds.), *Homology* (New York: John Wiley), Novartis Foundation Symposium, 222: 33–46.

Müller, G. B. & Newman, S. A., (2003), Origination of Organismal Form: The Forgotten Cause in Evolutionary Theory, in G. B. Muller & S. A. Newman (eds.), *Origination of Organismal*

Form: Beyond the Gene in Developmental and Evolutionary Biology (Cambridge, MA: The MIT Press), 3–12.

Nagy, L. M. & Williams, T. A., (2001), Comparative Limb Development as a Tool for Understanding the Evolutionary Diversification of Limbs in Arthropods: Challenging the Modularity Paradigm, in G Wagner (ed.), *The Character Concept in Evolutionary Biology* (New York: Academic Press), 455–488.

Nielsen, C. & Martinez, P., (2003), Patterns of Gene Expression: Homology or Homocracy? *Development, Genes, and Evolution* 213: 149–154.

Nijhout, H. F., (1990), Metaphors and the Role of Genes in Development, *BioEssays* 12: 441–446.

Norell, M. A., Novacek, M. J., (1992), The Fossil Record and Evolution: Comparing Cladistic and Paleontologic Evidence for Vertebrate History, *Science*, March 27, 1992, 255: 1690–1693.

Nusslein-Volhard, C. & Wieschaus, E., (1980), Mutations Affecting Segment Number and Polarity in Drosophila, *Nature* 287: 795–801.

Ohno, S., (1996), The Notion of the Cambrian Pananamalia Genome, *Proceedings of the National Academy of Sciences*, August 1996, 93: 8475–8478.

Olson, E. C., (1981), The Problem of Missing Links: Today and Yesterday, *Quarterly Review of Biology*, December 1981, 56: 405–441.

Orr, H. A., (1996–1997), Darwin v. Intelligent Design (Again), *Boston Review*, December 1996/January 1997, 29.

Padian, K., (1989), The Whole Real Guts of Evolution? *Paleobiology* 15: 73–78.

Panganiban, G. & Rubenstein, J. L. R., (2002), Developmental Functions of the Distal-less/Dlx Homeobox Genes, *Development* 129: 4371–4386.

Paulay, G., (1994), Biodiversity on Oceanic Islands: Its Origin and Extinction, *American Zoologist* 34: 134–144.

Peifer, M. & Bender, W., (1986), The Anterobithorax and Bithorax Mutations of the Bithorax Complex, *EMBO Journal* 5: 2293–2303.

Pennisi, E., (1997), Haeckel's Embryos: Fraud Rediscovered, *Science* 277: 1435.

Penny, D. & Phillips, M. J., (2004), The Rise of Birds and Mammals: Are Microevolutionary Processes Sufficient for Macroevolution? *Trends in Ecology and Evolution*, October 2004, 19: 516–522.

Peterson, S. N. & Fraser, C., (2003), The Complexity of Simplicity, *Genome Biology* 2: 1–8.

Rasmussen, A-S. & Arnason, U., (1999), Molecular Studies Suggest that Cartilaginous Fishes Have a Terminal Position in the Piscine Tree, *Proceedings of the National Academy of Sciences USA* 96: 2177–2182.

Raup, D. M., (1979), Conflicts Between Darwin and Paleontology, *Field Museum of Natural History Bulletin* 50: 22–29.

Raven, P. H. & Johnson, G. B., (2002), *Biology* (New York: McGraw Hill), 6th edition, 1229.

Reynolds, M. G., (2000), Compensatory Evolution in Rifampin-resistant Escherichia coli, *Genetics*, December 2000, 156: 1471–1481.

Richardson, M., (1998), Haeckel, Embyros, and Evolution, *Science* 280: 983–986.

Richardson, M. K., Hanken, J., Gooneratne, M. L., Pieau, C., Raynaud, A., Selwood, L. & Wright, G. M., (1997), There Is No Highly Conserved Embryonic Stage in the Vertebrates: Implications for Current Theories of Evolution and Development. *Anatomy and Embryology* 196: 91–106.

Rivlin, P. K., Gong, A., Schneiderman, A. M. & Booker, R., (2001) The Role of Ultrabithorax in the Patterning of Adult Thoracic Muscles in Drosophila melanogaster, *Dev Genes Evol* 211: 55–66.

Rokas, A., King, N., Finnerty, J. & Carroll, S. B., (2003), Conflicting Phylogenetic Signals at the Base of the Metazoan Tree, *Evolution and Development* 5: 346–359.

Romer, A. S. & Parsons, Thomas S., (1977), *The Vertebrate Body* (Philadelphia: W. B. Saunders).

Saier, M. H., (2004), Evolution of Bacterial Type III Protein Secretion Systems, *Trends in Microbiology* 12: 113–115.

Salazar-Ciudad, I., (2006), On the Origins of Morphological Disparity and Its Diverse Developmental Basis, *BioEssays*, November 2006, 28: 1112–1122.

Sander, K. & Schmidt-Ott, U., (2004), Evo-Devo Aspects of Classical and Molecular Data in a Historical Perspective, *Journal of Experimental Zoology B (Molecular and Developmental Evolution)* 302: 69–91.

Santos, M. A. S. & Tuite, M. F., (2004), Extant Variations in the Genetic Code, in L. R. de Pouplana (ed.), *The Genetic Code and the Origin of Life* (New York: Kluwer Academic and Plenum Publishers), 183–200.

Sapp, J., (1987), *Beyond the Gene* (New York: Oxford University Press).

Schwabe, C., (2002), Genomic Potential Hypothesis of Evolution: A Concept of Biogenesis in Habitable Spaces of the Universe, *The Anatomical Record* 268: 171–179.

Schwabe, C., (1994), Theoretical Limitations of Molecular Phylogenetics and the Evolution of Relaxins, *Comparative Biochemistry and Physiology* 107B: 167–177.

Schwabe, C. & Warr, G. W., (1984), A Polyphyletic View of Evolution: The Genetic Potential Hypothesis, *Perspectives in Biology and Medicine* 27: 465–485.

Scott, E. C., (2001), Evolution: Fatally Flawed Iconoclasm, *Science*, June 22, 2001, 2257–2258.

Sedgwick, A., (1894), On the Law of Development Commonly known as von Baer's Law; and on the Significance of Ancestral Rudiments in Embryonic Development, *Quarterly Journal of Microscopical Science* 36: 35–52.

Sermonti, G. & Fondi, R., (1980), *Dopo Darwin: Critica al' Evoluzionismo* (Milan: Rusconi).

Siew, N. & Fischer, D., (2003a), Analysis of Singleton ORFans in Fully Sequenced Microbial Genomes, *Proteins: Structure, Function, and Genetics* 53: 241–251.

Siew, N. & Fischer, D., (2003b), Twenty Thousand ORFan Microbial Protein Families for the Biologist? *Structure*, January 2003, 11: 7–9.

Simmons, N. B., (2005), An Eocene Big Bang for Bats, *Science*, January 28, 2005, 307: 527–528.

Stanley, S. M., (1982), Macroevolution and the Fossil Record, *Evolution* 36: 460–473.

Swift, D., (2002), *Evolution Under the Microscope: A Scientific Critique of the Theory of Evolution* (Stirling, Scotland: Leighton Academic Press).

Syvanen, M., (2002a), On the Occurrence of Horizontal Gene Transfer Among an Arbitrarily Chosen Group of 26 Genes, *Journal of Molecular Evolution* 54: 258–266.

Syvanen, M., (2002b), Recent Emergence of the Modern Genetic Code: A Proposal, *Trends in Genetics* 18: 245–248.

Telford, M. J. & Budd, G. E., (2003), The Place of Phylogeny and Cladistics in Evo-Devo Research, *International Journal of Developmental Biology* 47: 479–490.

Thompson, D. W., (1992), *On Growth and Form* (New York: Dover Publications).

Thomson, K. S., (1992), Macroevolution: The Morphological Problem, *American Zoologist* 32: 106–112.

Valentine, J. W., (2004), *On the Origin of Phyla* (Chicago: University of Chicago Press).

Valentine, J. W., (1992), The Macroevolution of Phyla, in J. H. Lipps & P. W. Signor (eds.), *Origin and Early Evolution of the Metazoa* (New York: Plenum Press).

Valentine, J. W. & Jablonski, D., (2003), Morphological and Developmental Macroevolution: A Paleontological Perspective, *International Journal of Developmental Biology* 47: 517–522.

Valentine, J. W. & Jablonski, D. & Erwin, D. H., (1999), Fossils, Molecules and Embryos: New Perspectives on the Cambrian Explosion, *Development* 126: 851–859.

Wake, D. B., (1999), Homoplasy, Homology and the Problem of 'Sameness' in Biology, in G. R. Bock and G. Cardew, (eds.), *Homology,* (New York: John Wiley), Novartis Foundation Symposium, 222: 24–33.

Wake, D., Homology—No. 222, CIBA Foundation Symposia Series, Novartis Foundation Symposium; Brian Hall, Dalhousie Univ., Halifax, Canada, 45.

Wassermann, G. D., (1978), Testability of the Role of Natural Selection Within Theories of Population Genetics and Evolution, *British Journal for the Philosophy of Science* 29: 223–242.

Webster, G. & Goodwin, B. C., (1982), The origin of species: a structuralist approach, *Journal of Social and Biological Structures* 5: 15–47.

Wegener, H. C., (2003), Ending the Use of Antimicrobial Growth Promoters Is Making a Difference, *ASM News* 69: 443.

Wells, J., (2003), *Icons of Evolution: Science or Myth?* (Washington, D. C.: Regnery Press).

Wills, M. A., (2002), The Tree of Life and the Rock of Ages: Are We Getting Better at Estimating Phylogeny? *BioEssays* 24: 203–207.

Woese, C., (2004), A New Biology for a New Century, *Microbiology and Molecular Biology Reviews* 68: 173–186.

Woese, C., (2002), On the Evolution of Cells, *Proceedings of the National Academy of Sciences* 99: 8742– 8747.

Wolf, Y. I., Rogozin, I. B. & Koonin, E. V., (2004), Coelomata and Not Ecdysozoa: Evidence from Genome-wide Phylogenetic Analysis, *Genome Research* 14: 29–36.

Yockey, H., (1992), *Information Theory and Molecular Biology* (Cambridge, England: Cambridge University Press).

Zakon, H. H., (2002), Convergent Evolution on the Molecular Level, *Brain, Behavior and Evolution* 59: 250–261.

Zimmer, C., (1995), Back to the Sea, *Discover* 16: 82–84.

page

3 Book photo: Natural History Museum (London)

4 Figure i:1: Getty Images; Car photos: Brian Gage

5 Car Photo: Brian Gage; Figure i:2: Photo courtesy of the University of Oklahoma History of Science Collections

6 Figure i:3: Historical Illustration, Public Domain

7 Photo composite: Brian Gage

8 Photo: Brian Gage

10 Photo: Brian Gage

15 Photo: J.Y. Chen

16 Illustration: J.Y. Chen

17 Figure 1:1: iStock Photo; Figure 1:2: Illustrations by Janet Moneymaker, except horse illustration, by Lucy Wells; Darwin Illustrations: Janet Moneymaker

19 Darwin Illustration: Janet Moneymaker

20 Figure 1:3: iStock Photo; Figure 1:4: Photo by Ken Lucas/ Visuals Unlimited

21 Illustration: Janet Moneymaker (from Kemp)

22 Bat Fossil Photo: Tom & Therisa Stack/Tom Stack & Associates

23 Photos: Brian Gage

24 Photo: Brian Gage

25 Fossil Nautilus Shell Photo: Alex Kerstitch/Visuals Unlimited; Modern Nautilus Shell Photo: iStock Photo; Fossil Shell Cut Away: Stock Photo; Modern Shell Cut Away: iStock Photo; Fossil Ginko Leaf Photo: Brian Gage; Modern Ginkgo Leaves Photo: iStock Photo; Fossil Comb Jelly Photo: Chen, *et al.;* Modern Comb Jelly Photo: Tom Stack Photo.

28 Illustrations: Janet Moneymaker

29 Illustrations: Janet Moneymaker (from Kemp)

30 Photos: iStock Photo

31 Illustration: Brian Gage

32 Illustration: Andrew Johnson; Figure 1:9: Photo by J.Y. Chen

34 Illustrations: Janet Moneymaker

35 Illustration: Brian Gage

36 Illustration: Brian Gage

39 Intro Text Photo: Brian Gage; Figure 2:1: Illustration courtesy of Jody Sjorgen

40 Photo: Brian Gage; Illustration: Jody Sjogren

41 Illustration: Jody Sjogren

43 Plane photo: Brian Gage; Refueling truck photo: iStock Photo

44 Wasp and Fly Illustrations: Tim Doherty; Figure 2:2: Illustrations by Janet Moneymaker

45 Figure 2:3: Illustrations by Janet Moneymaker; Figure 2:4: Mole Cricket Photo by Thomas Walker; Mole Photo by Charles Schurch Lewallen

46 Photo: Brian Gage

47 Photo: Brian Gage

49 Photos: iStock Photo

51 Illustration: Greg Neumayer

52 Intro Text Illustration: Greg Neumayer

57 Illustration: Tim Doherty

61 Photos: iStock Photo

65 Photo: Brian Gage

66 Photo: Brian Gage

67 Figure 4:1: Historical Illustration, Public Domain

68 Photos: Brian Gage

69 Figure 4:2: Photos used by permission from Michael Richardson, *et. al.* "Haeckel, Embryos, and Evolution," *Science* 280 (15 May 1998), pages 983-986

70 Photos: iStock Photo

72 Photos: Getty Images

73 Photos: Courtesy of NASA

74 Photo: iStock Photo; Photo composite: Brian Gage

75 Photos: iStock Photo; Figure 5:1: Photo courtesy of NASA

76 © www.micktravels.com

77 Photo: iStock Photo; Figure 5:2: Illustration courtesy of Jody Sjorgen

79 Photo: iStock Photo

83 Photo: Brian Gage

84 Photos: Brian Gage

85 Photo: Brian Gage

86 Photo: Brian Gage

87 Illustration: Janet Moneymaker

88 Figure 6:2: Illustration by Jody Sjorgen; Figure 6:3: The authors acknowledge the kind permission of the President and the Fellows of Wolfson College, Oxford to reproduce photographs held in the Kettlewell Archive.

90 Photo: Brian Gage

91 Photos: Brian Gage

92 Figure 6:5: Illustration by Janet Moneymaker

97 Illustration: Tim Doherty

98 Photo: Brian Gage

99 Photo: Science VU/ Visuals Unlimited

101 Illustrations: Tim Doherty

102 Illustrations: Tim Doherty

107 Photo: © Science VU/Visuals Unlimited

108 Photos: iStock Photo

110 Figure 7:1: DNA Illustration: Tim Doherty; Fish and Fish Eye Photos: iStock Photo

111 Figure 7:2: DNA Illustration: Tim Doherty; Fish and Fish Eye Photos: iStock Photo; Circuit Board and Transistor Photo: Mike Keas; CD Player Photo: iStock Photo

115 Illustration: Tim Doherty

116 Intro Text Illustration: Tim Doherty

117 Figure 8:2: Illustration by Tim Doherty

118 Illustration: Calvin Carl

119 Photo: Brian Gage

120 Photos: Brian Gage; Figure 8:3: Photo Courtesy of NASA; Figure 8:4: Illustration by Tim Doherty

121 Photos: iStock Photo

126 Photo: Brian Gage

127 Photo: iStock Photo

128 Photo: Brian Gage

130 Illustrations: Janet Moneymaker

131 Illustrations: Jonathan Moneymaker and Elena Wilken

134 Photos: Brian Gage

135 Illustrations: Janet Moneymaker

136 Illustrations: Janet Moneymaker

137 Illustrations: Janet Moneymaker (from Denton)

138 Photos: Brian Gage

142 Photo: Courtesy of the University of Oklahoma History of Science Collections

Cover Design and Inside Layout by Brian Gage Design
Brian Gage, Art Director/Designer
Calvin Carl, Designer/Production
Elena Wilken, Designer/Production
Marc Niedlinger, Designer/Production

INDEX

Words shown in bold type appear in the glossary.

About the Authors

Stephen C. Meyer earned his Ph.D. in the Philosophy of Science from Cambridge University for a dissertation on the history of origin-of-life biology, the logical structure of Darwin's argument and the methodology of the historical sciences. He also holds degrees in Physics and Geology. He is currently the Director and Senior Fellow of the Center for Science and Culture at the Discovery Institute in Seattle. Previously he worked as a professor at Whitworth College and a geophysicist with the Atlantic Richfield Company. He has co-authored or edited two other books: *Darwinism, Design, and Public Education* (Michigan State University Press) and *Science and Evidence of Design in the Universe.* He has authored articles in scientific journals such as the *Proceedings of the Biological Society of Washington and Trends in Ecology and Evolution* and in scientific books published with Cambridge University Press and Wessex Institute of Technology Press. He has also written many editorials on scientific topics for publications such as *USA Today*, *The Wall Street Journal* and *The Los Angeles Times.*

Scott Minnich holds a Ph.D. from Iowa State University. He is currently associate professor of microbiology at the University of Idaho. Previously, Dr. Minnich was an assistant professor at Tulane University. In addition, he did postdoctoral research with Austin Newton at Princeton University and with Arthur Aronson at Purdue University. Dr. Minnich's research interests are temperature regulation of *Y. enterocolitca* gene expression and coordinate reciprocal expression of flagellar and virulence genes.

Scott Minnich is widely published in technical journals including *Journal of Bacteriology, Molecular Microbiology, Journal of Molecular Biology, Proceedings of the National Academy of Sciences, Journal of Microbiological Method, Food Technology*, and the *Journal of Food Protection.*

Jonathan Moneymaker is a freelance technical writer, specializing in making complex topics easy for the non-expert to understand. His work has been used in classes and workshops at Boeing, Ford Motor Company, People Management International, and Walt Disney World Company (Disney University). Usually preferring to work in the background, his credited work includes the lighthearted look at the origins debate, *What's Darwin Got To Do With It?* His whimsical touch, analytical skill, and desire for clarity and readability all played key roles in writing this book.

Paul A. Nelson is a philosopher of science who received his Ph.D. from the University of Chicago (1998), where he specialized in the philosophy of biology and evolutionary theory. His dissertation, "Common Descent, Generative Entrenchment, and the Epistemology of Evolutionary Inference," critically evaluates the theory of common descent. He is currently a Fellow of the Discovery Institute, an Adjunct Professor at Biola University, and a member of the Society for Developmental Biology and the International Society for the History, Philosophy, and Social Studies of Biology. He has published articles in such journals as *Biology & Philosophy, Zygon,* and *Rhetoric and Public Affairs,* and scientific and philosophical papers in technical anthologies from MIT Press and Michigan State University Press.

Ralph Seelke received his undergraduate education at Clemson University (BS, Microbiology, 1973). He received his Ph.D. in Microbiology from the University of Minnesota and the Mayo Graduate School of Medicine in 1981, was a postdoctoral researcher at the Mayo Clinic until 1983, and has been an Associate Professor or Professor in the Department of Biology and Earth Sciences at the University of Wisconsin-Superior since 1989.

Dr. Seelke's research in experimental evolution has been well-regarded, and in 2004 he was a Visiting Scholar in the Department of Microbiology and Immunology at the Stanford University Medical School, conducting research to further our understanding of evolution. An authority on evolution's capabilities and limitations in producing new functions in bacteria, he has co-author eight publications in such journals as *Proceedings of the National Academy of Science, Journal of Bacteriology,* and *Molecular and General Genetics.* Prof. Seelke is a member of the American Society for Microbiology, and the American Scientific Affiliation.